LAB MA...
to Accompany

AGRISCIENCE

Fundamentals & Applications

Third Edition

Delmar is proud
to support FFA activities

Join us on the web at

www.Agriscience.Delmar.com

LAB MANUAL
to Accompany

AGRISCIENCE

Fundamentals & Applications

Third Edition

Walter York
Elmer L. Cooper
L. DeVere Burton

DELMAR
THOMSON LEARNING

Australia Canada Mexico Singapore Spain United Kingdom United States

**Lab Manual to Accompany Agriscience:
Fundamentals and Applications, Third Edition**
by
Walter York
Elmer L. Cooper
L. DeVere Burton

COPYRIGHT © 2002, 2004 by Delmar, a division of Thomson Learning, Inc.
Thomson Learning™ is a trademark used herein under license

Printed in United States
6 XXX 04

For more information contact Delmar,
3 Columbia Circle, PO Box 15015,
Albany, NY 12212-5015.

Or find us on the World Wide Web at
http://www.delmar.com

ALL RIGHTS RESERVED. No part of this work covered by the copyright hereon may be reproduced or used in any form or by any means—graphic, electronic, or mechanical, including photocopying, recording, taping, Web distribution or information storage and retrieval systems—without written permission of the publisher.

For permission to use material from this text or product, contact us by
Tel (800) 730-2214
Fax (800) 730-2215
www.thomsonrights.com

Library of Congress Catalog Number: 00-022680

ISBN 0-7668-1666-4

NOTICE TO THE READER

Publisher does not warrant or guarantee any of the products described herein or perform any independent analysis in connection with any of the product information contained herein. Publisher does not assume, and expressly disclaims, any obligation to obtain and include information other than that provided to it by the manufacturer.

The reader is expressly warned to consider and adopt all safety precautions that might be indicated by the activities herein and to avoid all potential hazards. By following the instructions contained herein, the reader willingly assumes all risks in connection with such instructions.

The Publisher makes no representation or warranties of any kind, including but not limited to, the warranties of fitness for particular purpose or merchantability, nor are any such representations implied with respect to the material set forth herein, and the publisher takes no responsibility with respect to such material. The publisher shall not be liable for any special, consequential, or exemplary damages resulting, in whole or part, from the readers' use of, or reliance upon, this material.

Table of Contents

Preface vii

SECTION 1 AGRISCIENCE IN THE INFORMATION AGE

    Exercise 1   The Scientific Method 2
    Exercise 2   Using a Microscope 5
    Exercise 3   The Cell 10
    Exercise 4   Maintaining a Healthy Environment 15

SECTION 2 YOU AND THE NEW MILLENNIUM

    Exercise 5   Your Career in Agriscience 22
    Exercise 6   Experience in Agriscience 31
    Exercise 7   Leadership in Agriscience 35
    Exercise 8   The Computer: An Information Tool 41

SECTION 3 NATURAL RESOURCES MANAGEMENT

    Exercise 9   The Effects of Air Pollution on Plants 48
    Exercise 10   Maintaining Clean Water 51
    Exercise 11   Preventing Soil Erosion 53
    Exercise 12   Soil Fertility 57
    Exercise 13   Growing Plants without Soil 61
    Exercise 14   Examining Our Forest Resources 64
    Exercise 15   Examining Our Wildlife Resources 67
    Exercise 16   Fish Anatomy: External and Internal 69

SECTION 4 INTEGRATED PEST MANAGEMENT

    Exercise 17   Insect Anatomy 78
    Exercise 18   Regulating Plant Growth via Chemicals 81
    Exercise 19   Examining Ecosystems 85

SECTION 5 PLANT SCIENCES

    Exercise 20   Stem Anatomy 88
    Exercise 21   Leaf Anatomy 92
    Exercise 22   Examining Plant Flowers 95

Exercise 23	Gas Production in Photosynthesis		99
Exercise 24	Water Movement in Plants		101
Exercise 25	Root Anatomy		103
Exercise 26	Seed Anatomy		107
Exercise 27	Requirements for Seed Germination		110
Exercise 28	Plant Propagation		114

SECTION 6 CROP SCIENCE

Exercise 29	Soil Organisms and Humus		118
Exercise 30	Enriching Soil through Decomposition		122
Exercise 31	Growing and Transplanting Vegetable Seedlings		125
Exercise 32	Grafting Fruit Trees		128
Exercise 33	Effects of Seed Planting Depths on Crop Production		130
Exercise 34	Storing Forage as Silage		133

SECTION 7 ORNAMENTAL USE OF PLANTS

Exercise 35	Responses of Plants to Light		136
Exercise 36	Cloning Plants for Uniformity		138

SECTION 8 ANIMAL SCIENCES

Exercise 37	Simple Digestion in Animals		142
Exercise 38	Digestion in Ruminant Animals		146
Exercise 39	Internal Parasites		151
Exercise 40	Controlling Diseases via Antibiotics		156
Exercise 41	Understanding Genetics		159
Exercise 42	Injection Procedures		163

SECTION 9 FOOD SCIENCE AND TECHNOLOGY

Exercise 43	Food Nutrients		168
Exercise 44	Food Preservatives		172
Exercise 45	Producing Dairy Products		174

SECTION 10 COMMUNICATIONS AND MANAGEMENT IN AGRISCIENCE

Exercise 46	Understanding Diminishing Returns		178
Exercise 47	Understanding Interest and Credit		183
Exercise 48	Preparing Balance Sheets		187
Exercise 49	Profit-Loss Statements		191

Preface

This laboratory manual was developed to accompany *Agriscience: Fundamentals & Applications*, Third edition. The topics and exercises throughout were designed to meet the science-based curriculum needs of the Agriscience 1 course in secondary schools. Although written with the organization of *Agriscience: Fundamentals & Applications*, Third edition in mind, this can be used as a stand alone laboratory manual.

EQUIPMENT AND SUPPLIES LIST

The equipment and supplies listed below were carefully selected to be used with the exercises in the *Lab Manual to Accompany Agriscience: Fundamentals & Applications*, Third edition. All items listed here are available from Carolina Biological Supply Company, 2700 York Road, Burlington, NC 27215. You may also place your order or request a catalog by calling toll free 1-800-334-5551.

Exercise 2 — **Carolina Biological Supply Catalog Number**
Compound microscope — 59–1244
Beginner's Slide Set, which includes a letter "e," newsprint, colored threads, cork, cornstarch, salt crystals, dust, cotton fibers, nylon fibers, silk fibers, wool fibers, and volcanic ash slides

Exercise 3
Compound microscope — 59–1244
Animal cell slide — 23575
Plant cell slide — B1066
Chromosomes—Mitosis slide — B551
Mitochondria slide — B554
Golgi apparatus slide — H335
Onion epidermis slide — B600A
Glass microscope slides — 63–2950
Coverslips — 63–2960
Medicine droppers — 736898
Pond water — 16–3380

Exercise 4
Scissors — 64–4775
Spray bottle — 66–5565
Magnifying Glass — 60–2204
Gloves — 19–9816

Exercise 5
Pen or marker — 64–4287
Scissors — 64–4770

	Carolina Biological Supply Catalog Number
Exercise 9	
Environmental Kit—Demonstrating Air Pollution	65–3090
Personal Safety Set	19–9998
Exercise 10	
Pond water	16–3380
Exercise 11	
Plant tray	62–9380
Sprinkling can	66–5572
Beaker	72–1507
Trowel	66–5210
Stopwatch	69–6913
Sod (1 square yard)	19–9999
Exercise 12	
6-inch flower pots with drainage outlets	66–5780
Vermiculite or pearlite	15–9722
Pea seeds	15–8862
Tomato seeds	15–9100
Nitrogen source	84–4112
Phosphorus source	85–2310
Potassium source	88–2900
Calcium source	85–1750
Materials for labeling pots	66–5950
Laboratory balance of scale	70–2152
Ruler or yardstick	70–2620
Exercise 14	
VCR or TV monitor	49–9075
Exercise 15	
VCR or TV monitor	49–9075
Exercise 16	
Dissection Lab Perch Kit	P3775D
Scissors	62–3005
Scalpel	62–5920
Sharp probe	62–7224
Magnifying glass	60–2204
Personal Safety Set	19–9998
Exercise 17	
Dissection Manual Grasshopper Kit	P3770
Fine dissection scissors	62–1810
Sharp probe	62–7405
Dissecting microscope	59–1822
Personal Safety Set	19–9998
Exercise 18	
100-ml beaker	72–1207
Razor blade	62–6930

	Carolina Biological Supply Catalog Number
Microscope slides	63–2950
Coverslips	63–2960
Toothpicks	64–4200
Forceps	62–4020
Personal Safety Set	19–9998

Exercise 19
Personal Safety Set	19–9998

Exercise 20
Monocot stem slide	B570
Dicot stem slide	B590
Microscope	59–1244
Set of colored pencils	64–4285

Exercise 21
Leaf section slide	B637
Compound microscope	59–1244
Set of colored pencils	64–4285

Exercise 22
Taraxacum young bud slide	B674
Taraxacum older bud slide	B770
Mature anther slide	B686
Pollen germination slide	B692
Compound microscope	59–1244
Set of colored pencils	64–4285

Exercise 23
Calomba Plant	16–2020
Manometer	68–2250
Large beaker	72–1517
Wax pencil	65–7730

Exercise 24
Osmosis and Diffusion Kit	68–4260

Exercise 25
Monocot root slide	B518B
Dicot root slide	B518A
Compound microscope	59–1244
Set of colored pencils	64–4285

Exercise 26
Seed of a plant slide	B71B
Microscope	59–1822
Fresh bean seeds	15–8302
Magnifying lens	60–2204
Scalpel	62–5920

	Carolina Biological Supply Catalog Number
Exercise 27	
Aluminum foil	71–3210
Beaker	72–1227
Seed-gro Kit	15–8172
Exercise 28	
Plant Propagation Kit	15–9601
Wandering Jew plant	15–7590
Spider or airplane plant	15–7535
Exercise 29	
What's in the Soil II—Microorganism Kit	84–1158
Compound microscope	59–1244
400-ml beaker	72–1210
Test tube rack	19–9330
Test tubes	19–8915A
Bunsen burner	21–6110
Personal Safety Set	19–9998

Exercise 31
Tomato seeds
Seed starting media
Flats
Spray bottle with water
Trowel or spoon
Plastic wrap
Large containers (10 oz. or larger)
Plant growing media
Liquid fertilizer
Ruler

Exercise 32
Knife
Grafting wax
Grafting tape
Apple or other fruit stock—
   ¼- to ½-inch diameter
Scions—¼- to ½-inch diameter

Exercise 33	
Glass jar—Pint size	71–5461
Soil	15–9705
Corn seeds	15–9242
Exercise 34	
1-quart jar with lid	19–9260
Hammer	64–6050
Exercise 36	
Plant II Cloning: Propagation of African Violets Kit	19–1109
African violet plant	15–7343
Bunsen burner	21–6110

	Carolina Biological Supply Catalog Number
Scalpel	62–5920
Water—Sterile	19–8697
Large beaker	72–1517
African violet potting soil	15–9700
Ethyl alcohol	86–1263
Hot plate	70–1015
Small plastic pots	66–5773
Personal Safety Set	19–9998

Exercise 37

Digestive tract of a pig or fetal pig	P1905C
Rubber gloves	19–9817
Scalpel	62–5920
Dissecting microscope	59–1822
Personal Safety Set	19–9998

Exercise 38

Rubber gloves	19–9817
Scalpel	62–5920
Dissecting microscope	59–1822
Compound microscope	59–1244
Personal Safety Set	19–9998

Exercise 39

Sheep Liver Fluke Microscope Slide Set	PS1230
Sheep Tapeworm Microscope Slide Set	PS1795
Pig Roundworm Microscope Slide Set	Z1035
Compound microscope	59–1244

Exercise 40

Marking pen	64–4287

Exercise 41

Calculator

Exercise 42

Two or three 10-ml syringes
½-inch needles
Three containers of water
Food dye—red, green, blue
Three to four grapefruit
Ruler

Exercise 43

Test tubes	19–8915A
Test tube rack	19–9340
Eyedropper	73–6898
Hot plate	70–1015
Sudan IV	89–2963
Buiret reagent	84–8211
Potassium iodide—Iodine solution	86–9051
Benedict's solution	84–7111
Distilled water	85–7201

Exercise 44
	Carolina Biological Supply Catalog Number
Hot plate	70–1015
20-ml test tube	19–8915A
250-ml beaker	72–1209
Test tube holder	70–2900
Sulfuric acid solution	89–3411
Funnel	73–4012
Filter paper	71–2742
Spatula	70–2742
Ethyl alcohol	86–1281

Exercise 45
12-inch piece of cheesecloth	71–2690
Hot plate	70–1015
Cooking thermometer	74–5384
Rennin or rennilase	20–2375

Exercise 46
Calculator	91–2442

Exercise 47
Calculator	91–2442

Exercise 48
Calculator	91–2442

Exercise 49
Calculator	91–2442

About the Author

Walter York, the author of this laboratory manual, is the Agriscience Instructor and Principal at Chapel Hill High School. He was most recently Agriculture 2 + 2 Project Director at Northeast Texas Community College. He was previously an Agriculture Occupational Education Specialist with the Texas Education Agency in Austin. A former high-school agriculture instructor, he received Bachelor of Science and Master of Science degrees from Texas A&M University. He was the Outstanding Teacher of Vocational Agriculture in Texas in 1986 and recognized by the Texas Forestry Association as the Outstanding Forestry Teacher in 1987. He holds all active and honorary degrees awarded by the National FFA Association.

SECTION: 1

Agriscience in the Information Age

EXERCISE 1: THE SCIENTIFIC METHOD

Materials Needed
- Textbook
- Other materials deemed appropriate by students

Purpose To introduce students to the scientific method and give them practice using it.

Procedure

1 Study the steps of the scientific method as identified in the following. These steps will help you throughout this laboratory manual.

   a. Step 1. Identify the problem. Exactly what do you want to find out? Limit your topic to a single research objective.

   b. Step 2. Review the literature. Read up on and become familiar with the topic.

   c. Step 3. Form a hypothesis. Develop a hypothesis or statement to be proven or disproven.

   d. Step 4. Prepare a project proposal. Outline how the project should be done. Include timelines, facilities, equipment required, costs, and a description of how you will do the project.

   e. Step 5. Design the experiment. Develop a plan for carrying out the project.

   f. Step 6. Collect the data. Conduct the experiment. Record what you measure or observe.

   g. Step 7. Draw conclusions. Summarize the results. Make calculations. Determine if the information collected allows you to accept or reject the hypothesis.

   h. Step 8. Prepare a written report. The written report provides you with a permanent record of your research.

2 Using the scientific method in agriscience

   a. Step 1. Identify the problem. The following problem has been identified for you. Complete the remaining steps of the scientific method.

   How does the color of a material affect the heat and light absorption or reflection?

Observations

1. Step 2. Review the literature. List any references that you used for review in the space provided. Include the name of the references and other pertinent information.

2. Step 3. Form a hypothesis. Write your hypothesis in the space provided.

3. Step 4. Prepare a project proposal. Outline your project in the space provided. Include timelines, facilities, equipment required, costs, and a description of how you will do the project.

4. Step 5. Design the experiment. Develop your plan to test your hypothesis in the space provided.

5. Step 6. Collect the data. Record any data in the space provided Design your own data format.

6. Step 7. Draw conclusions. Summarize your results in the space provided. Confirm or reject your hypothesis.

_____
_____
_____
_____
_____
_____
_____
_____

7. Step 8. Prepare a written report. Using a computer and word processing program, prepare a written report not to exceed two pages explaining your project and the results.

Conclusions

1. How did the scientific method help you to prove or disprove your hypothesis?

_____
_____
_____
_____

2. Which step in the scientific method do you consider to be the most important and why?

_____
_____
_____
_____

3. Which step in the scientific method do you consider to be the least important and why?

_____
_____
_____
_____

EXERCISE 2:

USING A MICROSCOPE

Materials Needed

✔ Compound Microscope

✔ Beginner's Slide Set, which includes a letter "e," newsprint, colored threads, cork, cornstarch, salt crystals, dust, cotton fibers, nylon fibers, silk fibers, wool fibers, and volcanic ash slides

Purpose To introduce the compound microscope and provide practice in using it.

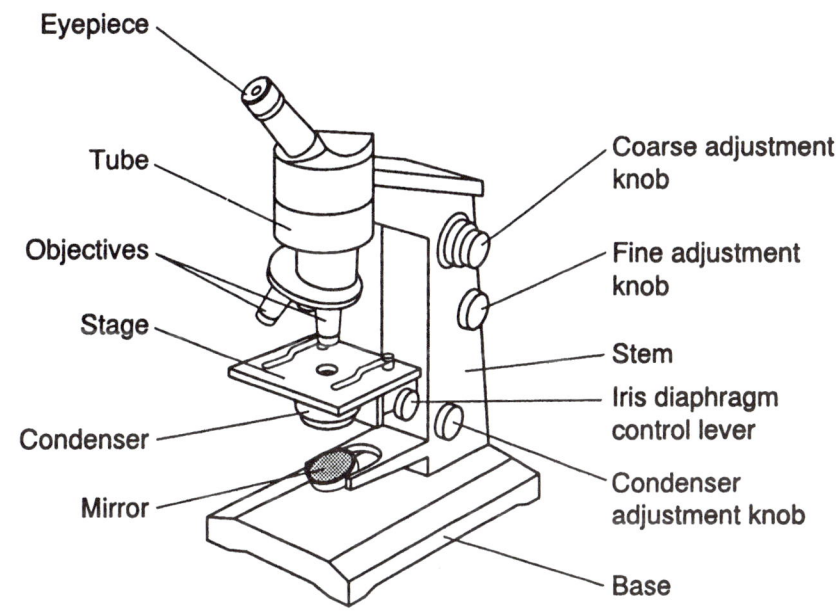

Procedure

1 Study the preceding diagram of a microscope. Learn the names and functions of the parts.

  a. **Eyepiece** — magnifying lens at the top of the microscope

  b. **Tube** — connects eyepiece to the objectives

  c. **Objectives** — lens that magnifies the object to be viewed

  d. **Stage** — platform that holds the slide for viewing

  e. **Condenser** — concentrates the light source onto the object being observed

  f. **Light source** — provides light to view the slide on the stage; it may be a mirror or a lamp.

  g. **Coarse-adjustment knob** — knob used for initial focusing

  h. **Fine-adjustment knob** — knob used for final focusing

  i. **Stem** — connects the base to the lens assembly

  j. **Iris-diaphragm control lever** — controls the amount of light to the slide

  k. **Condenser adjustment knob** — adjusts light brilliance through the condenser

l. **Base** — platform to which entire microscope is attached; sits on table or desk

② Prepare the microscope and view the slide.

a. If your microscope has a mirror, position the mirror so that light is reflected through the hole on the stage. If your microscope has a light, turn it on.

b. Place the letter "e" microscope slide on the stage and secure it with the stage clips. Make sure the coverslip faces up. Center the slide over the stage hole.

c. Rotate the low-power objective so that it locks into position directly over the specimen and over the stage hole.

d. Look at the stage and use the coarse-adjustment knob to bring the objective down to within ¼ inch of the coverslip.

e. While observing through the eyepiece, rotate the coarse-adjustment knob counterclockwise to bring the image into view. Use the fine-adjustment knob for sharp focusing. **Remember:** *Always focus upwards.*

f. Practice opening and closing the iris diaphragm by moving the lever on the side of the condenser clockwise to provide more light and counterclockwise to provide less light. More light is needed for higher-power objectives. Less light is required for transparent objects. Record your observations in Observation 2.

g. While looking through the eyepiece, observe the letter "e" slide while moving the slide forward, backward, left, and right. Describe what happens to the "e" in Observation 3.

h. Carefully turn the objective to the high-power lens. Look through the eyepiece at the letter "e" slide. If it is not in focus, adjust the fine-adjustment knob only. (Never use the coarse-adjustment knob to adjust the high-power objective lens because this lens is very close to the slide and could break it.)

i. Observe each of the other slides found in the beginner's slide set. Draw a picture of or describe what you see for each slide in Observation 4 that follows.

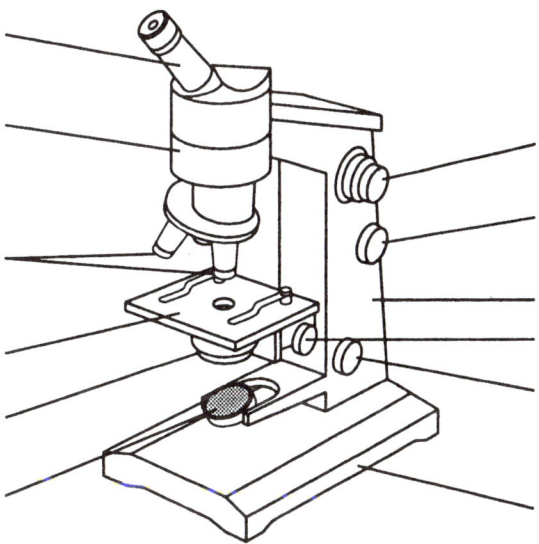

Observations

1. Label the parts of the microscope.

2. Describe the effects of opening and closing the diaphragm.

3. Describe the appearance of the letter "e" under the microscope and how it moves as the slide moves.

4. Draw a picture of or describe in detail each of the following as it appears under the microscope.

   a. Newspaper print

   b. Colored threads

c. Cork

d. Cornstarch

e. Salt crystals

f. Dust

g. Cotton fibers

h. Nylon fibers

i. Silk fibers

j. Wool fibers

k. Volcanic ash

Conclusions

1. How does the image you see through the microscope compare with the nonmagnified object seen directly?

2. Explain how the movement of the slide on the stage and the movement as viewed through the microscope differ.

3. In what areas of agricultural research would the microscope be useful? List at least five.

EXERCISE 3: THE CELL

Materials Needed
✔ Compound microscope
✔ Animal cell slide
✔ Plant cell slide
✔ Chromosomes—Mitosis Slide
✔ Mitochondria slide
✔ Golgi apparatus slide
✔ Onion epidermis slide
✔ Glass microscope slides
✔ Coverslips
✔ Medicine droppers
✔ Pond water

Purpose To study and compare the structures of plant and animal cells, and to identify the common parts of cells.

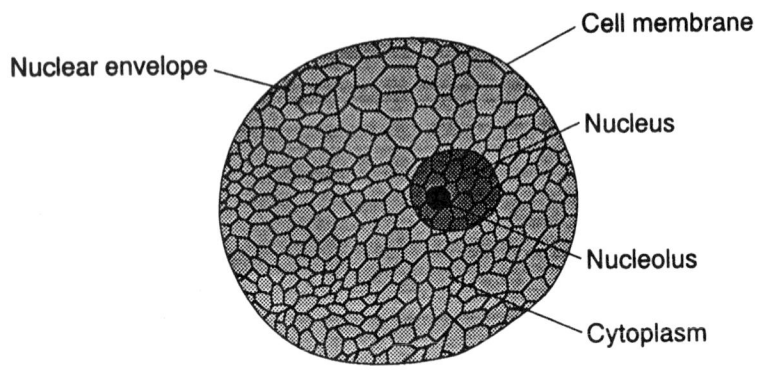

Animal Cell

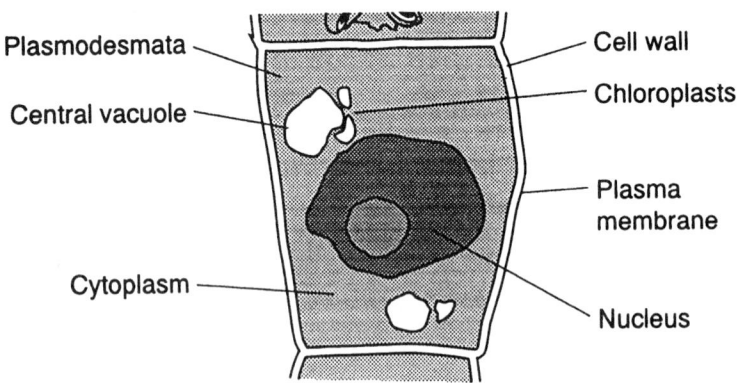

Plant Cell

Procedure

1 Viewing animal cells

a. Place the general animal cell slide on the stage of the microscope and view it under low power.

b. Switch to high power. In Observation 1 that follows, identify these parts: cell membrane, cytoplasm, nucleus, nucleolus, and nuclear envelope.

c. Place the chromosome slide on the stage of the microscope and view it under low power.

d. Switch to high power. In Observation 2, identify the chromosomes in the drawing.

e. Place the mitochondria slide on the stage of the microscope and view it under low power.

f. Switch to high power. In Observation 3, identify the mitochondria in the drawing.

g. Place the Golgi apparatus slide on the stage of the microscope and view it under low power.

h. Switch to high power. In Observation 4, identify the Golgi apparatus in the drawing.

i. Use the general animal cell picture to help you complete the statements in Observation 5.

❷ Viewing plant cells

a. Place the general plant cell slide on the stage of the microscope and view it under low power.

b. Switch to high power. In Observation 6, identify the following parts: cell wall, plasma membrane, cytoplasm, nucleus, central vacuole, and plasmodesmata.

c. Use the general plant cell picture to help you complete the statements in Observation 7.

❸ Viewing onion cells

a. Place the onion epidermis slide on the stage of the microscope and view it under low power.

b. Switch to high power. In Observation 8, draw an onion cell and identify the following parts: cell wall, plasma membrane, cytoplasm, nucleus, and vacuoles.

❹ Viewing pond water

a. Using a medicine dropper, place a drop of pond water on the center of a clean slide.

b. Place a coverslip over the pond water.

c. View the slide under low power to see the larger organisms. Switch to high power to see the smaller organisms.

d. In Observation 9, draw at least three different organisms or cells you observe under the microscope.

e. With your instructor's help, try to identify the organisms that you observed.

12 ■ SECTION 1: AGRISCIENCE IN THE INFORMATION AGE

Observations

1. Using the following drawing, identify the following:

   cell membrane
   nucleus
   nuclear membrane
   cytoplasm (cytosol)
   rough endoplasmic reticulum
   lipid droplets

   centrosomes
   chromosomes
   mitochondria
   Golgi Body (Apparatus)
   ribosome cluster

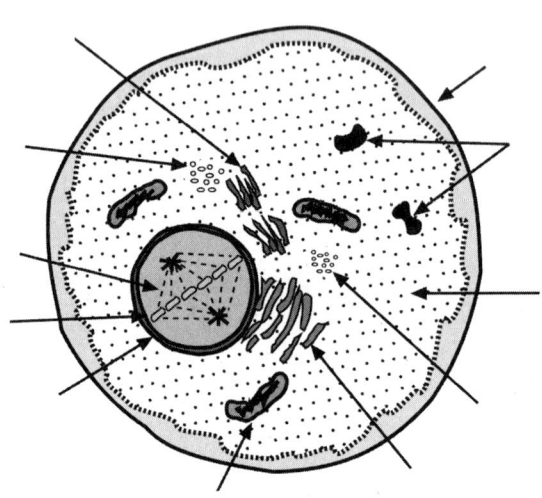

2. Fill in the blanks to make the following statements true:

   a. The _____ _____ is the outermost structure of an animal cell.

   b. The fluid medium in which much of the intermediary metabolism of the cell takes place is the _____.

   c. The large organelle that contains the cell's inherited information and serves to control and coordinate all metabolism is the _____.

   d. The _____ separates the nuclear contents from the rest of the cell.

   e. The _____ are tangled strands that extend throughout the nucleus and contain genes.

   f. Cellular organelles in which the chemical energy in food molecules is converted to a form useful to cellular metabolism are called _____.

   g. The _____ _____ plays a part in the storage or secretion of metabolic products.

   h. The site of ribosomal RNA synthesis and ribosome assembly is the _____.

3. Identify the parts of the general plant cell on the following diagram.

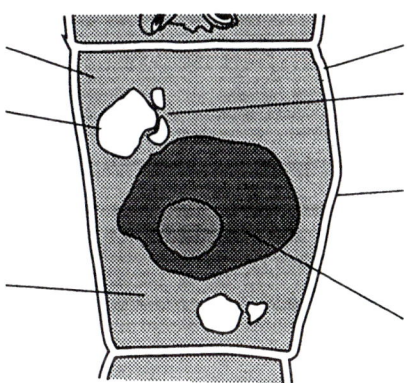

4. Fill in the blanks to make the following statements true.

   a. The _____ _____ is the outermost structure of a plant cell.

   b. The membrane-enclosed sac that occupies up to 90 percent of the cell volume is the _____.

   c. The large organelle that contains the cell's inherited information and serves to control and coordinate all metabolism is the _____.

   d. The _____ consists of a viscous fluid matrix that contains water, many organic and inorganic substances, and a variety of tiny, specialized, functional structures.

   e. The _____ forms the outer boundary of a cell and maintains the chemical composition of the cell by regulating the passage of substances in and out of the cell.

   f. _____ are organelles that often contain lamella; chloroplasts are an example.

5. Draw an onion cell and label its parts.

6. Draw at least three organisms found in pond water. Identify each organism, if possible.

14 ■ SECTION 1: AGRISCIENCE IN THE INFORMATION AGE

Conclusions

1. What parts are different between animal and plant cells?

2. What are the differences between the general plant cell and the onion cell?

3. What are some other plant cells that would be different from the general plant cell?

4. What parts of the general animal and plant cells are similar?

5. What differences can be observed in the general shapes of the animal and plant cells?

EXERCISE 4: MAINTAINING A HEALTHY ENVIRONMENT

Materials Needed

- Paper strips
- Dried plant materials such as leaves
- Scissors
- Spray bottle
- Water
- Three plastic cups
- Sand
- Topsoil
- Plastic
- Biodegradable plastic
- Cotton fibers
- Synthetic fibers
- Magnifying glass
- Gloves

Purpose To determine the solid-waste materials that are biodegradable and factors that affect the rate and biodegradation of solid-waste materials.

Procedure NOTE: Students may work individually or in teams of three.

1. Preparing samples
   a. Cut the paper into ½-by-2-inch strips. Cut the two kinds of plastic into ½-by-2-inch strips.
   b. Cut the synthetic and natural fibers into 2-inch lengths.
   c. Cut the dried plant material accordingly, depending on its size.
   d. Separate the materials into three identical piles having one of each material.

2. Preparing the landfill
   a. Label three plastic cups as follows:

   Cup #1 — Sand
   Cup #2 — Dry topsoil
   Cup #3 — Moist topsoil

   b. Fill cup #1 with sand, up to 1¼ inch from the top, packing the sand tightly.
   c. Fill cups #2 and #3 with topsoil, up to 1¼ inch from the top, packing the soil lightly.
   d. Water the topsoil in cup #3 using a spray bottle. Water until soil is moist but not excessively so that there is water at the bottom of the cup. The soil should be uniformly moist and not have any residual water pooled at the bottom of the cup.

3. Adding materials to the landfill
   a. Make initial observations of the original condition of the following materials to be tested; record your observations in Table 1.

   1. Paper strips
   2. Plastic strips
   3. Biodegradable plastic strips
   4. Natural fibers
   5. Synthetic fibers
   6. Dried plant material

   NOTE: Other materials may be tested for biodegradation

15

b. Test the original strength of the fiber (synthetic and natural) by stretching the fiber; record fiber strength using the key provided in Table 2.

c. Insert a piece of each of the materials to be tested into the soil of each cup, covering about half the surface area of each material. A portion of each material tested should be exposed above the soil surface. The exposed portions can be used to locate the buried materials and as reference guides to the original conditions of the materials.

d. Press the materials lightly into the soil and cover each cup with a clear plastic top.

e. Place your cups on a tray and store them at room temperature.

Observations

1. Record your observations of the original condition of the materials in Table 1.
   NOTE: Ask your instructor for multiple copies of Table 1 for use throughout the experiment.

2. Record your observations of the original strength of the fibers in Table 2.

3. Observe your set up once a week and record your results in Table 1. Remove the materials one by one, gently clean off the soil material and examine each carefully with a 10X magnifying glass. Look for differences in general characteristics, such as size, shape, texture, breaks in the fibers, and fading of colors.

4. Test the strength of the fibers (synthetic and natural) once a week and compare these results to those of the original fibers. Record your observations in Table 2.

Conclusions

1. Which of the soil materials provided for the best biodegradation? Explain you answer.

2. When did you first notice signs of degradation?

3. What difference was there in the degradation between the natural fiber and the synthetic fiber?

4. Which of the cups had the most degrading activity? Explain why you think it did.

5. Which materials decomposed most quickly?

6. Which materials did not show any decomposition?

Name _____ Date _____

Table 1 Cup # _____ Soil Type _____

Material	General Appearance and Condition	Color	Texture
Paper			
Plastic			
Biodegradable Plastic			
Natural Fiber			
Synthetic Fiber			
Dried Plant Material			

Name _____ Date _____

Table 2   Rate Fiber

Fiber	At Start	First Week	Second Week	Third Week	Fourth Week	Fifth Week	Sixth Week	Seventh Week
Natural (Cotton)	+++							
Synthetic (Polyester)	+++							

Rate fiber strength as follows:

+++   Original strength at start

++   Starting to show signs of disintegration, such as breaks or cuts of the thinner fibers that make up the sample

+   Very weak, but still held together

-      Disintegrated or decomposed

SECTION: 2

You and the New Millennium

EXERCISE 5:

YOUR CAREER IN AGRISCIENCE

Purpose To introduce the different agriscience careers.

Materials Needed

✔ Pencil

✔ Pen or marker

✔ Thumbtack

✔ Sheet of posterboard— 18 inches × 24 inches

✔ Scissors

✔ Glue

✔ Occupational Outlook Handbook

✔ Dictionary of Occupational Titles

Procedure

1 Preparing the career wheel

 a. Using a pair of scissors, cut out the career wheel.

 b. Cut out a piece of posterboard the same size as the career wheel.

 c. Glue the career wheel to the posterboard.

 d. Cut a rectangular piece of posterboard to measure 8½ inches by 11 inches.

 e. Place the thumbtack in the middle of the career wheel and fasten it to the 8½-by-11-inch piece of posterboard.

 f. Place a dark arrow pointing toward the wheel on the 8½-by-11-inch piece of posterboard. When you finish, your career wheel should look like the preceding diagram.

2 Choosing a career.

   a. Spin the career wheel. The arrow will point to one of the career areas.

   b. Select a job for the career area from the list found under Procedure 3.

   c. Using the *Occupational Outlook Handbook* and *Dictionary of Occupational Titles* complete Observation 1 for the job you selected.

   d. Spin the wheel again and repeat Procedures 2b and 2c to complete Observation 2.

   e. Without using the wheel, select a job of your choice from the lists under Procedure 3. Complete Observation 3 for the job you selected.

3 Agricultural career areas and jobs listed for each:

   a. Production Agriculture

Crop Farmer	Vegetable Farmer
Fruit and Nut Farmer	Horticultural Grower
Livestock Producer	Poultry Producer
Other List _____	

   b. Agriscience Processing, Products, and Distribution

Ag Establishment Inspector	Butcher
Cattle Buyer	Christmas Tree Grader
Cotton Grader	Farm Stand Operator
Federal Grain Inspector	Food and Drug Inspector
Food Processing Supervisor	Fruit Distributor
Fruit and Vegetable Grader	Flower Grader
Fruit Press Operator	Grain Buyer
Grain Broker	Hog Buyer
Grain Elevator Operator	Livestock Yard Supervisor
Livestock Commission Agent	Meatcutter
Meat Inspector	Produce Buyer
Milk Plant Supervisor	Quality Control Supervisor
Produce Commission Agent	Weights and Measures Official
Tobacco Buyer	Winery Supervisor
Other List _____	Wool Buyer

   c. Horticulture

Floral Designer	Floral Shop Operator
Florist	Golf Course Superintendent
Greenhouse Manager	Greenskeeper
Horticulturist	Hydroponics Grower
Landscape Architect	Landscaper
Nursery Operator	Plant Breeder
Turf Farmer	Turf Manager
Other List _____	

d. Forestry

Forester
Heavy Equipment Operator
Logging Operations Inspector
Nursery Operator
Plant Breeder
Tree Surgeon

Forest Ranger
Log Grader
Lumber Mill Operator
Park Ranger
Timber Manager
Other List _____

e. Renewable Natural Resources

Animal Behaviorist
Animal Ecologist
Animal Taxonomist
Environmental Conservation
    Officer
Environmentalist
Fire Warden
Trapper
Forest Ranger
Game Farm Supervisor

Game Warden
Ground Water Geologist
Park Ranger
Range Conservationist
Resource Manager
Soil Conservationist
Forest Firefighter/Warden
Water Resources Manager
Wildlife Manager
Other List _____

f. Agriscience Supplies and Service

Aerial Crop Duster
Ag Chemical Dealer
Animal Groomer
Animal Inspector
Animal Trainer
Artificial Breeding
    Technician
Biostatistician
Chemical Distributor
Computer Analyst
Computer Programmer
Custom Operator
Dog Groomer
Farm Appraiser
Farrier
Feed Ration Developer
    and Analyst
Fiber Technologist
Field Sales Representative,
    Animal Health Products
Field Sales Representative,
    Crop Chemicals, Machinery
Kennel Operator
Meteorological Analyst
Pet Shop Operator
Poultry Hatchery Manager
Poultry Inseminator
Salesperson
Sheep Shearer

Ag Aviator
Ag Equipment Dealer
Animal Health Products
    Distributor
Animal Keeper
Artificial Breeding
    Distributor
Artificial Inseminator
Chemical Applicator
Computer Operator
Computer Salesperson
Dairy Management Specialist
Farm Auctioneer
Feed Mill Operator
Fertilizer Plant Supervisor
Field Inspector
Field Sales Representative,
    Agricultural Equipment
Harness Maker
Harvest Contractor
Horse Trainer
Insect and Disease Inspector
Lab Technician
Pest Control Technician
Poultry Field Service
    Technician
Sales Manager
Service Technician
Other List _____

SECTION 2: YOU AND THE NEW MILLENNIUM ■ 25

g. **Agricultural Mechanics**

Ag Construction Engineer	Ag Electrician
Ag Equipment Designer	Ag Plumber
Ag Safety Engineer	Diesel Mechanic
Equipment Operator	Hydraulic Engineer
Irrigation Engineer	Land Surveyor
Machinist	Parts Manager
Research Engineer	Safety Inspector
Soil Engineer	Welder
Other List _____	

h. **Agriscience Professions**

Ag Accountant	Ag Advertising Executive
Ag Association Executive	Ag Consultant
Ag Corporation Executive	Ag Economist
Ag Educator	Ag Extension Agent
Ag Extension Specialist	Ag Journalist
Ag Lawyer	Ag Loan Officer
Ag Market Analyst	Ag Mechanics Teacher
Ag News Director	Agriculture Attaché
Agronomist	Animal Cytologist
Animal Geneticist	Animal Nutritionist
Animal Physiologist	Animal Scientist
Agriculturist	Avian Veterinarian
Bacteriologist	Biochemist
Bioengineer	Biophysicist
Botanist	Computer Specialist
Credit Analyst	Daily Nutrition Specialist
Dendrologist	Electronic Editor
Embryologist	Environmental Educator
Entomologist	Equine Dentist
4-H Youth Assistant	Farm Appraiser
Farm Broadcaster	Farm Investment Manager
Food Chemist	Foreign Affairs Official
Graphic Designer	Herpetologist
Horticulture Instructor	Hydrologist
Ichthyologist	Information Director
International Specialist	Invertebrate Zoologist
Land Bank Branch Manager	Liminologist
Magazine Writer	Mammalogist
Marine Biologist	Marketing Analyst
Media Buyer	Microbiologist
Mycobiologist	Nematologist
Organic Chemist	Ornithologist
Ova Transplant Specialist	Paleobiologist
Parasitologist	Pharmaceutical Chemist
Photographer	Plant Cytologist
Plant Ecologist	Plant Geneticist
Plant Nutritionist	Plant Pathologist
Plant Taxonomist	Pomologist
Poultry Scientist	Public Relations Manager
Publicist	Publisher
Reproductive Physiologist	Rural Sociologist
Satellite Technician	Scientific Artist
Scientific Writer	Silviculturist
Software Reviewer	Soil Scientist
Vertebrate Zoologist	Veterinarian
Veterinary Assistant	Veterinarian Pathologist
Veterinarian Technician	Virologist
Vidiculturist	Vocational Agriculture
Other List _____	Instructor/FFA Advisor

Observations

1. Find the information requested in the Observation 1 by using the *Dictionary of Occupational Titles* and the *Occupational Outlook Handbook*.

2. Find the information requested in the Observation 2 by using the *Dictionary of Occupational Titles* and the *Occupational Outlook Handbook*.

3. Find the information requested in the Observation 3 by using the *Dictionary of Occupational Titles* and the *Occupational Outlook Handbook*.

Conclusions

1. How does the nature of work differ for each of the three jobs? How is it similar?

2. How do the working conditions differ for each of the three jobs? How are the conditions the same?

3. How do the qualifications for each of the three jobs differ? How are the qualifications the same?

4. How does the job outlook compare for each of the three jobs?

5. How do the earnings compare for each of the three jobs?

OBSERVATION 1

Career Wheel Challenge

Career Area [ ]   **Job Selected** [ ]

Nature of the Work:

_____
_____
_____
_____
_____

Working Conditions:

_____
_____
_____
_____
_____

Training and Other Qualifications:

_____
_____
_____
_____
_____

Job Outlook:

_____
_____
_____
_____
_____

Earnings:

_____
_____
_____
_____
_____

OBSERVATION 2

Career Wheel Challenge

Career Area _____   **Job Selected** _____

Nature of the Work:
_____
_____
_____
_____

Working Conditions:
_____
_____
_____
_____

Training and Other Qualifications:
_____
_____
_____
_____

Job Outlook:
_____
_____
_____
_____

Earnings:
_____
_____
_____
_____

OBSERVATION 3

Career Wheel Challenge

Career Area _____   **Job Selected** _____

Nature of the Work:

_____
_____
_____
_____
_____

Working Conditions:

_____
_____
_____
_____
_____

Training and Other Qualifications:

_____
_____
_____
_____
_____

Job Outlook:

_____
_____
_____
_____
_____

Earnings:

_____
_____
_____
_____
_____

Cut out the following career wheel and glue it to the posterboard cut to the same size. Use the thumbtack to attach this to the 8½-by-11-inch piece of posterboard. Spin the wheel to complete the exercise.

Career Wheel

- Horticulture
- Agricultural Mechanics
- Agriscience Professions
- Agriscience Supplies and Services
- Forestry
- Renewable Natural Resources
- Production Agriculture
- Agriscience Processing Products and Distribution

EXERCISE 6: EXPERIENCE IN AGRISCIENCE

Materials Needed
- ✔ Official FFA Manual
- ✔ SAEP Record Book
- ✔ Pen or pencil

Purpose To help in planning a Supervised Agricultural Experience Program (SAEP).

Procedure

1 Planning your SAEP

a. List the various agricultural crop or livestock enterprises produced successfully in your community (Observation 1). Circle those that interest you.

b. List the various agribusinesses found in your community that support production agriculture (Observation 2). Circle those that interest you.

c. List improvement activities you could carry out (Observation 3).

d. List agriscience skills you could perform (Observation 4).

e. Look through your SAEP record book. What type of information would you record in it (Observation 5)?

f. Complete the resources inventory (Observation 6).

g. Complete the "Selecting a Supervised Agricultural Experience Program" form (Observation 7).

h. Optional: Visit the SAEP of an outstanding agriscience junior or senior or an American FFA Degree recipient (if one is nearby). Ask how he or she got started.

2 SAEP awards

a. Use your *Official FFA Manual* to identify individual awards for which you could apply based on your proposed SAEP. List these in Observation 8.

b. Ask your agriscience instructor to help you list other state and local awards for which you can apply (Observation 9).

Observations

1. Crop and livestock enterprises produced in my community: _____

2. Agribusinesses found in my community: _____

3. Improvement projects (applied activities) that I can do: _____

4. Supplementary skills (applied activities) that I can perform: _____

5. Information that I would record in my SAEP record book: _____

6. Complete the Resources Inventory.

7. Complete the "Selecting a Supervised Agricultural Experience Program."

8. National individual awards for which I can apply: _____

9. State or local awards for which I can apply: _____

RESOURCES INVENTORY

1. Name _____ Age _____ Class _____

2. Address _____ Phone _____

3. Parents' names _____

4. Occupations _____

5. Number of brothers and sisters _____

6. I live on a farm _____ in town _____ on an acreage _____

7. Is land available for you to rent to produce crops? Yes _____ No _____

    a. If yes, how many acres? _____ b. Which crops? _____

    c. Location of the land _____

8. Are facilities available for you to rent to produce livestock or livestock products?

    Yes _____ No _____

    a. If yes, what type? _____

    b. Number _____ c. Location of facilities _____

9. Do you have space available for a garden? Yes _____ No _____

10. Do you have a greenhouse available for your use? Yes _____ No _____

11. Do you have facilities for doing mechanical work? Yes _____ No _____

12. Would you be interested in producing livestock on the school farm?

    Yes _____ No _____ If yes, what type of livestock? _____

13. Would you be interested in producing crops on the school farm?

    Yes _____ No _____ If yes, what crop? _____

14. Do you have an agricultural job available to you?

    Yes _____ No _____ Describe _____

SELECTING A SUPERVISED AGRICULTURAL EXPERIENCE PROGRAM

for

_____
Name of Student

Instructions: Use this form to tentatively decide on a beginning agriscience SAEP. This information will be used in agriscience classes to develop detailed plans for obtaining experiences.

My interest in agriscience is _____
_____
_____

Based on my interest and the opportunities available to me to get practical experience in agriscience, I plan to include the following in my agriscience SAEP:

1. Production or productive enterprises (examples: beef, dairy, nursery production, Christmas trees)
   _____
   _____

2. Placement in an agribusiness (examples: supply store, florist shop, nursery, golf course, landscape contracting)
   _____
   _____

3. Improvement (examples: landscape your home, fertilize your lawn, plant trees)
   _____
   _____

4. Agriscience skills (examples: change spark plugs, weld, change the oil in small engines)
   _____
   _____

5. Other activities (examples: projects on school facility)
   _____
   _____

EXERCISE 7: LEADERSHIP IN AGRISCIENCE

Materials Needed
- ✔ Official FFA Manual
- ✔ Parliamentary Guide for FFA

  or

- ✔ Robert's Rules of Order
- ✔ Pen or pencil

Purpose To introduce the procedures for conducting FFA meetings and the rules of parliamentary law.

Procedure

1 Conducting FFA meetings

a. Study the following diagram of the FFA meeting room. Learn where each of the officers is stationed and the symbol of each office.

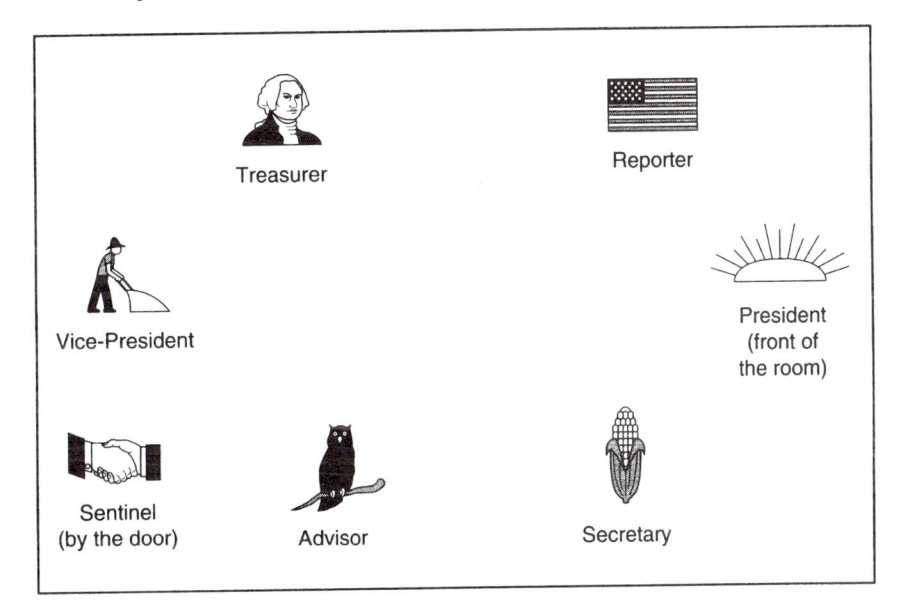

```
        Treasurer              Reporter

                                   President
                                   (front of
Vice-President                     the room)


  Sentinel
(by the door)    Advisor        Secretary
```

b. In the *official FFA Manual*, read over the opening and closing ceremonies for all FFA meetings.

c. Let each class member read a part in the ceremonies. Take turns. Go over the ceremonies several times.

d. Select one officer's position and write the part in the ceremonies (Observation 2).

2 Learning parliamentary law

a. In the *Official FFA Manual*, study the order of business for a meeting. Complete Observation 3.

b. Study the classifications, names, purposes, and pertinent information for the primary motions used in conducting FFA business. Use the *Parliamentary Guide for FFA* or *Robert's Rules of Order*. Complete Observation 4.

c. Practice conducting a meeting using the sample business found in Observation 5. Fill in the blanks appropriately.

36 ■ SECTION 2: YOU AND THE NEW MILLENNIUM

Observations

1. Complete the following meeting room diagram by listing the officers and the symbols where they should be stationed.

```
┌─────────────────────────────────────────────┐
│                                             │
│                                             │
│                                             │
│                                   Front of  │
│                                   the       │
│                                   Room      │
│                                             │
│                                             │
│       Door                                  │
│                                             │
└─────────────────────────────────────────────┘
```

2. Write one part in the opening ceremony. _____

3. List the order of business for an FFA meeting. _____
   _____
   _____
   _____
   _____
   _____
   _____
   _____
   _____

4. Complete the blanks in the following chart.

Name of Motion	Kind of Motion	Debatable (Yes or No)	Amendable (Yes or No)	Vote Required (Majority Two-thirds or None)	Second Required (Yes or No)
Main Motion					
Lay on the Table					
Division of the House					
Amend					
Take from the Table					
Refer to a Committee					
Previous Question					
Reconsider					
Parliamentary Inquiry					
Rescind					
Suspend the Rules					
Adjourn					

5. List the information requested for each motion. _____
   _____
   _____
   _____
   _____

Practicing Parliamentary Law

1. A member moves to paint the interior of the agriculture building. Write an introductory statement; include proper terminology. _____

2. The motion is debated by two members. Write discussion for and against the motion.

3. Lay the amended motion on the table. Write an introductory statement; include proper terminology. _____

4. A motion is made to have the annual parent-member banquet during National FFA Week in February. Write an introductory statement, include proper terminology.

5. A member moves to refer the motion to a committee. Write an introductory statement; include proper terminology.

6. A member discusses the motion to refer to a committee. Write discussion for the motion to refer.

7. A member moves the previous question. Write the proper terminology to make this motion.

8. A member moves to adjourn the meeting. Write the proper terminology for this motion.

Conclusions

1. How can learning leadership in agriscience help you to become a better citizen in your community?

2. In what other meetings or organizations can you use parliamentary law?

3. List five individuals in your community that you consider to be leaders. Give reasons for selecting each.

EXERCISE 8:

THE COMPUTER: AN INFORMATION TOOL

Materials Needed

- ✔ Microcomputer — IBM or Compatible or Macintosh
- ✔ FFA Proficiency Award computer program
- ✔ Printer
- ✔ Printer paper

Purpose To introduce the microcomputer and explain its use.

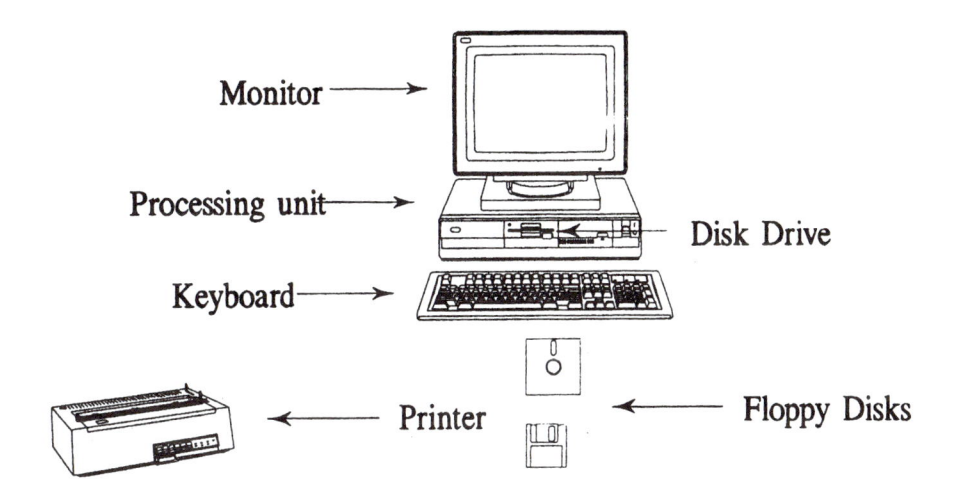

Procedure

1. Study the preceding drawing of a computer. Label the diagram in Observation 1 with the names of the following parts and learn their functions.

    a. **Disk drive** — rotates a floppy disk at high speeds and allows information to be stored on the disk or read by the computer

    b. **Floppy disk** — that is made of flexible material encased in a square cover and that stores information. May be 3½ inches or 5¼ inches in size

    c. **Processing unit** — includes the central processing unit (CPU), main memory, and their associated parts

    d. **Monitor** — screen used to read the computer's output; also referred to as a cathode ray tube (CRT)

    e. **Keyboard** — "typewriter" used to put information into the computer

    f. **Printer** — puts on paper the information found on a disk. Printed output is often referred to as a *hard copy*

2 Booting the computer

a. Turn on the monitor.

b. If your computer has a hard drive, the disk operating system (DOS) is probably installed here and will "boot up." Turn your processing unit on.

c. If your computer has only a floppy drive, put the disk operating system (DOS) disk into drive A. Turn on the processing unit.

d. Get the A> to appear on the screen.

3 Running the program

a. Remove the DOS disk from drive A and insert the FFA Proficiency Award application disk.

b. At the A>, type **PROF.**
(**NOTE:** Your instructor should have entered the chapter name and number prior to using the disk.)

c. If two drives are on the computer, insert a formatted data disk in drive B. If the computer has only one drive, have the data disk handy.

d. Press **enter** to continue.

e. Select **setup** from the main menu. Select **configure** from the submenu.

f. Configure the disk to correspond to the number of drives and type of monitor that your computer has. Press **escape** to save. If you have only one disk drive, you will need to swap out the program disk and the data disk when instructed from this point on.

g. Select **setup** from the main menu a second time. Select **initialize** from the submenu. Follow the instructions on the screen to initialize the data disk.

h. Select **setup** from the main menu a third time. Select **add** from the submenu. From the list of proficiency awards, select the one that you would like to work toward achieving. Enter the number. The computer will create the application on your data disk. Press **enter** to continue.

i. Next, enter the beginning date, beginning inventory, and ending date for the first year of your application. Your instructor can help you with this information, if needed. If the information entered is correct, type **Y**.

j. Select **edit** from the main menu and select the number of the proficiency award on which you wish to work. A page appears on the screen that tells you the last date that you worked on each page of the application. Press **enter**.

k. On this screen, select **1** for page 1 of the application.

l. Complete Observation 2.

m. Enter the information from Observation 2 on cover of the computer application (when applicable). Press **escape** to save cover.

4 Printing the completed page

a. Press **escape** to return to the main menu. Select **setup** from the main menu.

b. Select **printer** from the submenu. From the list of printers, select the printer that you have. Enter the number and press **enter**.

c. From the main menu, select **print**. From the submenu, select **print**. Select the number of the proficiency that you wish to print.

d. Select **1** from the submenu for cover of the application.

e. Turn on your printer. Adjust the paper in the printer if necessary. Press **enter** to begin printing.

f. Select **quit** from the main menu.

Observations

1. Label the parts of the computer in the following diagram.

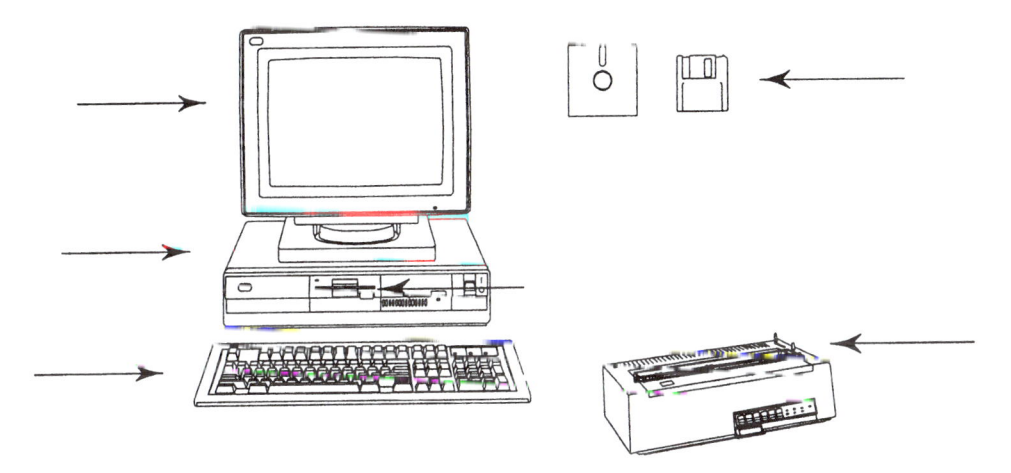

2. Complete the following information sheet.

State in which you live _____

Name _____ Age _____

Date of birth _____ Social Security Number _____

Address _____

Telephone (include area code) _____

Parent/guardian name _____

Parent/guardian occupation _____

Name of high school _____

School address _____

School telephone (include area code) _____

Chapter advisor(s) _____

Year FFA membership began _____ *

Year elected to greenhand degree _____ *

Year elected to chapter FFA degree _____ *

Year elected to state FFA degree _____ *

Years agricultural education completed _____ *

*You may want to put in target dates for these blanks. Ask your agriscience instructor for assistance.

Conclusions

1. How can a computer serve as an information tool for agricultural workers?

2. Other than applying for awards, what are some other programs and other uses for the computer?

3. What are some agricultural jobs that could make use of the computer as an information tool? List five.

SECTION: 3

Natural Resources Management

EXERCISE 9:

THE EFFECTS OF AIR POLLUTION ON PLANTS

Purpose   To determine the effects of polluted air on plant production.

Materials Needed

- Environmental Kit—Demonstrating Air Pollution
- Personal Safety Set

Procedure

1. Follow the directions in the kit's student guide for growing the squash and marigold plants.

2. Set up the environmental chamber and the control chamber according to the directions found in section II of the student guide in the kit.

3. Set up the sulfur dioxide (SO_2) generator according to the instructions found in section III of the student guide.

4. Generate the SO_2 in the generator as described in section IV of the student guide. (Do so only under direct supervision of your instructor.)

5. Remove the domes from the two germination chambers containing the squash and marigold seedlings.

6. Examine the plants in each chamber daily, twice daily if possible.

7. Compare the plants in the environmental chamber to the corresponding plants in the control chamber.

8. Record the dates observed and your observations for five consecutive days. Post this information to the tables in Observation 1.

SAFETY NOTE

Read all instructions carefully before proceeding. Follow your instructor's directions before proceeding with any step.

Observations

1. Complete the tables on the next page with the observation dates and observations from your experiment.

Environmental Chamber Plants Plant Condition	Day 1 Date: _____	Day 2 Date: _____	Day 3 Date: _____	Day 4 Date: _____	Day 5 Date: _____
Color Changes in Stems or Leaves					
Loss of Rigidity (Dropping of Stems or Leaves)					
Differences in Root Hairs					
Other Differences					

Control Chamber Plants Plant Condition	Day 1 Date: _____	Day 2 Date: _____	Day 3 Date: _____	Day 4 Date: _____	Day 5 Date: _____
Color Changes in Stems or Leaves					
Loss of Rigidity (Dropping of Stems or Leaves)					
Differences in Root Hairs					
Other Differences					

Conclusions

1. What differences were observed between the squash and the marigold seedlings in the environmental chamber?

2. Why might one species be more affected than the other?

3. What are the differences between the plants in the control chamber and those in the environmental chamber?

4. What might eventually happen if air pollution were unchecked?

5. What effect does air pollution have on your life?

EXERCISE 10:

MAINTAINING CLEAN WATER

Materials Needed

- ✔ Total Coliform Kit
- ✔ Basic Equipment Set
- ✔ Well water
- ✔ Pond water
- ✔ Treated water

SAFETY NOTE

Do not touch any viable cultures with your fingers. Be sure to wash your hands after each experiment. In handling all cultures, use aseptic techniques and common sense.

Purpose   To determine the degree of pollution in water.

Procedure

1. Sterilize the sterifil apparatus, the Swinnex water filter, and the 2-dram vial and screw cap according to instructions included in the coliform test kit.

2. Collect an untreated, fresh well-water sample in the sterilized vial.

3. Prepare the sterifil apparatus with a 0.45-micron, 47-millimeter membrane filter according to the procedure listed in the coliform test kit.

4. Prepare the 47-millimeter petri dish with an absorbent pad and MF-Endo medium as outlined in the coliform test kit.

5. Attach the hand vacuum system to the sterifil apparatus as outlined in the coliform test kit.

6. After filtration of the test sample, release the vacuum and place the sample in the petri dish according to the procedures outlined in the coliform test kit.

7. Incubate the petri dish at room temperature for 48 hours.

8. Using a hand lens or stereomicroscope perform a colony count using one of the formulas listed in the coliform test kit. Record your results in the table in Observation 1.

9. Follow the procedures previously outlined and in the coliform test kit to perform the same test on pond water and treated water. Record the results in the tables in Observations 2 and 3.

10. Dispose of all materials according to your instructor's directions

52 ■ SECTION 3: NATURAL RESOURCES MANAGEMENT

Observations

1. Record the results of the untreated water test in the following table.

Untreated Fresh Water Sample	No. Coliform/100 ml	Desirable Level	Permissible Level

2. Record the results of the pond water test in the following table.

Pond Water Sample	No. Coliform/100 ml	Desirable Level	Permissible Level

3. Record the results of the treated water test in the following table.

Treated Water Sample	No. Coliform/100 ml	Desirable Level	Permissible Level

Conclusions

1. What can cause the differences in the results of the tests on the three water samples?

   _____
   _____
   _____
   _____

2. What could be the sources of the coliform detected in the water samples?

   _____
   _____
   _____
   _____

3. How do each of the samples compare to the desirable levels allowed by the EPA? To the permissible levels?

   _____
   _____
   _____
   _____

EXERCISE 11: PREVENTING SOIL EROSION

Purpose To compare the rates and amounts of erosion that result from different land uses.

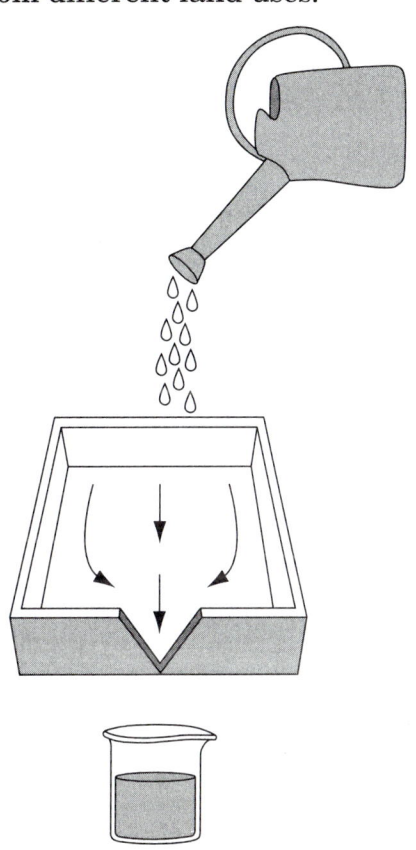

Materials Needed

- Six plant trays (or shoe boxes cut to 5 inches deep)
- Plastic sheeting to line trays
- Six sprinkling cans
- Six beakers
- Trowel
- Scissors
- Stopwatches
- Soil
- Sod (grass-covered soil from a lawn)
- Water
- Supports to form inclines for trays

Procedure

1 Preparing the materials

a. Cut a "V" notch in one end of each plant tray (or shoe box).

b. Line each tray with plastic, leaving an overhang of about 2 inches at the notch. This will serve as a spout.

c. Number and label each tray as you fill each of the six trays differently as follows:

1. moist soil packed firmly (bare soil)
2. sod (cover crop)
3. moist soil packed firmly; make packed furrows running across the tray's width (contouring)
4. moist soil packed firmly; make packed furrows running down the length of the tray (plowing up and down a hill)
5. moist soil packed firmly; form steps across the width of the tray with the trowel (terracing)

53

6. alternate three strips of moist packed soil with three strips of sod (strip cropping)

d. Position the trays on supports so the trays are tilted at an equal incline. Place beakers under the plastic overhangs at the "V" notches (spouts).

❷ Conducting the experiment

a. Measure an equal amount of water into each sprinkling can.

b. With the help of fellow students, hold the sprinkling cans about 1 foot over the high point of each tray. Pour water steadily for 5 seconds onto each tray simultaneously.

c. In Observation 1, record the amount of timed that water flows from the spout of each box.

d. After the water finishes draining, record the amount of runoff in each beaker in Observation 2.

e. Let the runoff settle, and then measure the volume of sediment in each beaker. Record the results in Observation 3.

f. Arrange the trays to increase the slope and repeat Procedures 2a through 2e, recording the results in Observations 4 through 6.

SECTION 3: NATURAL RESOURCES MANAGEMENT ■ 55

Observations

1.
	Tray 1	Tray 2	Tray 3	Tray 4	Tray 5	Tray 6
Amount of Time Water Flows from the Spout						

2.
	Beaker 1	Beaker 2	Beaker 3	Beaker 4	Beaker 5	Beaker 6
Amount of Runoff						

3.
	Beaker 1	Beaker 2	Beaker 3	Beaker 4	Beaker 5	Beaker 6
Volume of Sediment						

4.
	Tray 1	Tray 2	Tray 3	Tray 4	Tray 5	Tray 6
Amount of Time Water Flows from the Spout at Increased Slope						

5.
	Beaker 1	Beaker 2	Beaker 3	Beaker 4	Beaker 5	Beaker 6
Amount of Runoff at Increased Slope						

6.
	Beaker 1	Beaker 2	Beaker 3	Beaker 4	Beaker 5	Beaker 6
Volume of Sediment at Increased Slope						

Conclusions

1. Which tray had the most runoff? Which tray lost the most soil?

2. Which tray's content best prevented erosion?

3. Is there a relationship between the time it took for each tray to finish draining and how much soil washed off? Explain your answer.

4. How does the slope affect the soil loss?

5. How would you apply what you have learned from this experiment to soil conservation?

EXERCISE 12: SOIL FERTILITY

Materials Needed

- Seven identical 6-inch flower pots with drainage outlets at the bottom
- Pot chip or screen
- Vermiculite or pearlite
- Mixing bin or box
- Pea and tomato seeds
- Nitrogen source— NH_4NO_3
- Phosphorus source— superphosphate
- Potassium source— KCl
- Calcium source— $CaCO_3$
- Materials for labeling pots
- Laboratory balance or scale
- Ruler or yardstick

Purpose To understand the value of soil nutrients by observing the effects of a deficiency of certain nutrients in plants.

Procedure

1. Place a screen or pot clip over the hole in the bottom of each pot.

2. Fill each pot with soil to 1 inch from the top.

3. Label the pots to correspond with the names of the treatments in column one of the following chart. One pot should be labeled for each treatment.

Treatment	Nitrogen	Phosphorus	Potassium	Calcium
No Fertilizer	None	None	None	None
Complete Fertilizer	1.1 g NH_4NO_3	1.25 g Superphosphate	.3 g KCl	None
Complete Fertilizer with Calcium	1.1 g NH_4NO_3	1.25 g Superphosphate	.3 g KCl	1.5 g $CaCO_3$
Nitrogen Deficient	None	1.25 g Superphosphate	.3 g KCl	None
Phosphorus Deficient	1.1 g NH_4NO_3	None	.3 g KCl	None
Potassium Deficient	1.1 g NH_4NO_3	1.25 g Superphosphate	None	None
Calcium Only	None	None	None	1.5 g $CaCO_3$

4. Weigh the amounts of fertilizer specified in columns two, three, four, and five of the preceding chart. Dump the soil into a bin or box and mix thoroughly with the fertilizer before returning the soil to the pot. Be sure not to mix materials from separate pots. Wash your hands after handling fertilizer.

5. Divide each pot in half with an imaginary line. Plant five pea seeds in one half and five tomato seeds in the other half. Water them in.

6. Place the pots in a location with adequate light and temperature for growing. Ensure all pots have the same growing conditions.

57

7. Check the plants at regular intervals and water them as needed.

8. When the seeds have germinated and the plants are established, thin the plants so that each pot has the same number of each kind of plant.

9. Observe the plants each week, making note of differences in color, height, width, and any other differences between pots. Record this information in the table in Observation 1.

10. At the conclusion of the experiment, harvest all plants by pulling them up by the roots. Weigh all the plants of the same kind in each pot together and record the weights on the chart in Observation 2.

11. Let the plants dry in the sun for 4 days. Be sure to keep the plants from different pots separated. Weigh each kind of plant in each pot again separately. Record the dry weights on the chart in Observation 2.

SECTION 3: NATURAL RESOURCES MANAGEMENT ■ 59

Observations

1. In the following table record your observations of rate of growth, color, and so on.

				Treatment				
Observations	No Fertilizer	Complete Fertilizer	Complete Fertilizer with Calcium	Nitrogen Deficient	Phosphorus Deficient	Potassium Deficient	Calcium Only
Week 1								
Week 2								
Week 3								
Week 4								
Week 5								
Week 6								
Week 7								
Week 8								

60 ■ SECTION 3: NATURAL RESOURCES MANAGEMENT

2. In the following chart record the harvest weight for each kind of plant for each treatment.

	Green Weight		Dry Weight	
Treatment	Peas	Tomatoes	Peas	Tomatoes
No Fertilizer				
Complete Fertilizer				
Complete Fertilizer with Calcium				
Phosphorus Deficient				
Potassium Deficient				
Calcium Only				

Conclusions

1. What were the signs of deficiencies in the plants you grew, if any?

   a. Nitrogen _____

   b. Phosphorus _____

   c. Potassium _____

   d. Calcium _____

2. Which plants seemed to grow best in the soil to which calcium was added? _____

3. How did the green weights compare to the dry weights? _____

EXERCISE 13: GROWING PLANTS WITHOUT SOIL

Materials Needed

- Hydroponics Experiment Kit
- Personal Safety Set

SAFETY NOTE

Wear eye protection when handling any chemical solution. Handle chemical reagents with care and follow your instructor's directions. Avoid contact with skin or eyes.

Purpose — To demonstrate the art of growing plants without soil (hydroponics).

Procedure

1. Fill the transparent plastic culture chamber with deionized water to the ridge located near the top edge.

2. Add fifty drops of each of the following nutrient solutions to the water-filled chamber and stir with a clean glass rod.
   a. $Ca(NO_3)_2$ solution (calcium nitrate)
   b. KNO_3 solution (potassium nitrate)
   c. $MgSO_4$ and KH_2PO_4 solution (magnesium sulphate and potassium phosphate)

3. Add five drops of the iron chelate solution to the chamber.

4. Add five drops of the trace elements solution to the chamber.

5. Carefully place a seed inside the slit of a germinating plug. This should then be inserted into a hole provided in the white germinating float. When each of the twelve holes have been filled with seeds (and plugs), carefully place the float into the chamber so that all plugs are in contact with the liquid in the chamber.

6. Replace the opaquing shield on the lower portion of the culture chamber, if it has been removed. The shield will inhibit the growth of algae in the solution and should be removed to observe root growth.

7. Replace the vented cover and allow the plants to germinate. A south or west window is the best location for the culture chamber.

8. As the solution is used by the plants, it should be brought back to the original level by adding deionized water. After the seeds have germinated, replace the nutrients and trace elements once a week.

9. Observe the growth of the plants daily for two weeks and record the progress on the chart in Observation 1.

Observations

1. Record the daily progress of the plants in the following chart.

Day	Plant Growth Conditions			
	Length of Shoots and Roots	Coloration of Roots	Coloration of Leaves	Other Conditions
1				
2				
3				
4				
5				
6				
7				
8				
9				
10				
11				
12				
13				
14				

Conclusions

1. How can the plants germinate without soil?

2. Did you observe any deficiencies in the plants?

3. What color were the roots of the plants?

4. What possibilities do you see for the use of hydroponics in the future?

EXERCISE 14:

EXAMINING OUR FOREST RESOURCES

Materials Needed

- ✔ Pen or pencil
- ✔ VCR
- ✔ TV monitor
- ✔ Video — Fundamentals of Forestry
- ✔ Video — Tree Identification
- ✔ Video — Practice Tree Identification
- ✔ Video — Lumber Production

Purpose   To become familiar with the various resources in the forest and the uses of some of these resources.

Procedure

1 View the video entitled *Fundamentals of Forestry*. Identify the different parts of the forestry contest in Observation 1.

2 View the video entitled *Tree Identification*.

3 View the video entitled *Practice Tree Identification*. Record your identifications in Observation 2.

NOTE: If you live in an area where forests are available, you may want to practice identification on actual trees.

4 Complete Observations 3 through 5 while viewing the video entitled *Lumber Production*.

Observations

1. List the ten different areas of the forestry contest identified in the *Fundamentals of Forestry* video.

_____
_____
_____
_____
_____
_____
_____
_____
_____
_____
_____
_____
_____
_____
_____
_____
_____
_____

2. Record your tree identifications in the following chart.

Class 1	Class 2
1. _____	1. _____
2. _____	2. _____
3. _____	3. _____
4. _____	4. _____
5. _____	5. _____
6. _____	6. _____
7. _____	7. _____
8. _____	8. _____
9. _____	9. _____

Class 3	Class 4
1. _____	1. _____
2. _____	2. _____
3. _____	3. _____
4. _____	4. _____
5. _____	5. _____
6. _____	6. _____
7. _____	7. _____
8. _____	8. _____
9. _____	9. _____

66 ■ SECTION 3: NATURAL RESOURCES MANAGEMENT

3. Identify ten different jobs that you observed in the *Lumber Production* video.

4. Identify the harvesting and hauling equipment used in the *Lumber Production* video.

5. Identify five pieces of safety equipment seen in the *Lumber Production* video.

Conclusions

1. Describe how plywood is made.

2. Which parts of the forestry contest relate to the skills an actual forester would need?

3. What are some of the courses recommended in the *Fundamentals of Forestry* video for someone interested in a forestry career?

4. Make a list of things that you see or use every day that are forestry products.

EXERCISE 15:

EXAMINING OUR WILDLIFE RESOURCES

Materials Needed

- VCR
- TV monitor
- Video—Wildlife: An American Heritage
- Video—Future of Wildlife
- Pencil or pen

Purpose To become familiar with the various wildlife resources and how we can protect their futures.

Procedure

1. View the video *Wildlife: An American Heritage*. Complete Observations 1 through 3 as you watch the video.

2. View the video *Future of Wildlife*. Complete Observations 4 and 5 as you watch the video.

NOTE: Agriscience classes in areas where wildlife is abundant may wish to take field trips to survey the wildlife in their areas.

Observations

1. List at least ten animals shown in the video *Wildlife: An American Heritage*.

   _____
   _____
   _____
   _____

2. List two birds that are extinct.

   _____
   _____
   _____
   _____
   _____
   _____

3. List three species of birds that are rarely seen and are almost extinct.

   _____
   _____
   _____
   _____

68 ■ SECTION 3: NATURAL RESOURCES MANAGEMENT

4. List five species of animals or birds that have benefited from wildlife management.

5. What are five ways that you observed in the *Future of Wildlife* video that wildlife areas are being lost daily?

Conclusions

1. In what ways has money been collected by the government for use to acquire and manage wildlife habitats?

2. Seventy percent of the endangered species are endangered because of the loss of what?

3. What can you do as an individual to help maintain wildlife for the future?

EXERCISE 16:

FISH ANATOMY: EXTERNAL AND INTERNAL

Materials Needed

- Dissection Lab Perch Kit
- One pair scissors
- One scalpel
- One sharp probe
- Magnifying glass
- Personal Safety Set

SAFETY NOTE

Always wear eye protection and rubber gloves when handling dissection specimens. Wash hands before leaving the laboratory.

Purpose To become familiar with the major internal and external features and structures of the perch and with their functions.

Procedure

1 Study and become familiar with the following terms:
  a. **Dorsal**—back or upper
  b. **Ventral**—abdominal or lower
  c. **Anterior**—forward or front
  d. **Posterior**—hind or rear
  e. **Median**—towards the middle

2 Observe Figure 12–6 in *Agriscience: Fundamentals & Applications, 3E*, study the parts and learn their functions as defined.

External Anatomy

a. **Mouth**—at anterior end of the animal. The position of the mouth reflects the manner whereby the fish feeds. The fish possessing a terminal mouth feeds by overtaking prey while swimming. The fish possessing a dorsal mouth is usually a surface feeder. If the mouth is ventral, the fish is usually a bottom feeder.

b. **Nostril**—two pairs, each pair opens into an olfactory (smell) chamber and allows water to pass through the chamber. The movement of water is essential to odor perception.

c. **Eye**—two large spherical eyes are located on each side of the head

d. **Gill openings**—allow water passing through the gills to exit

e. **Operculum**—a semicircular flap that covers the gills

f. **Pectoral fins**—paired fins found on either side of the perch

g. **Pelvic fins**—paired fins found close together on the ventral surface of the fish trunk

h. **Anterior dorsal fin**—single fin found on the dorsal side of the fish and supported by 13 to 15 spiny rays

69

i. **Posterior dorsal fin**—single fin located directly behind the anterior dorsal fin and supported by seven to eight soft rays
j. **Caudal fin**—the fleshy end of the tail is surrounded by this fin; the type of caudal fin found on the perch is called *homocercal*
k. **Anal fin**—single fin located behind the anus on the ventral side and supported by soft rays
l. **Anus**—opening through which feces are eliminated from the fish body
m. **Skin**—envelope for the body and the first line of defense against disease; it also affords protection from, and adjustments to, environmental factors that influence life
n. **Mucous glands**—glands located in the skin, the secretions of which give the slimy touch and odor to the fish
o. **Fish scales**—thin, flexible plates that provide a flexible "armor" to protect the fish. They are used in growth and age studies. The scales of perch are called *ctenoid* because the exposed portion bears tiny, spiny structures called *cteni* (teeth)
p. **Lateral line system**—a sensory organ system restricted to the fishes and aquatic stages of amphibians; it enables a fish to detect localized disturbances from currents or vibrations caused by fixed or moving objects

❸ Study external anatomy
a. Place the perch on the dissection tray, side down, with the head facing to the left, dorsal fins up.
b. Locate and identify the external features of the perch. Label these on the diagram in Observation 1.
c. Using forceps, carefully remove a fish scale from the trunk region and examine it under a magnifying glass. Identify the cteni. Complete Observation 2.

❹ Study the internal parts of the perch found in text Figure 12–6. Each part's function is discussed in the appropriate section under Internal Anatomy.

Internal Anatomy

a. **Muscular system**—controls body movements
1. With a scalpel, carefully cut a large (2-inch) square "window" on the lateral surface of the perch directly below the anterior dorsal fin.

2. Using forceps, carefully peel away the skin to expose the muscle layer beneath. Identify and label the trunk muscles on the diagram in Observation 3. Trunk muscles are attached to the skeleton and are of primary use in producing body movement.

b. **Respiratory system**—serves in the exchange of gases with the environment

1. Using scissors, carefully cut the operculum at its base on the head to expose the gills. Observe the figure in the kit.
2. Extend the cut posteriorly from the base of the angle of the jaw around the gills. Raise and remove the operculum.
3. Pry open the mouth with a probe and observe the mouth and pharynx (pharyngeal cavity). Feeding is accomplished by sucking prey into the mouth, unlike those fishes that bite their prey.
4. Identify the following parts of the respiratory system on the fish and label them on the diagram in Observation 3.
    a) **Palate**—solid roof extending from the mouth cavity to the pharynx.
    b) **Gills**—organs of respiration that have thin walls that are supplied with blood capillaries, as are our lungs.
    c) **Gill arches**—serve as attachments for the gills.
    d) **Gill filaments**—tiny "fingers" running at right angles to the length of each gill. Oxygen dissolved in water diffuses through the thin wall of each gill filament into capillaries, and carbon dioxide diffuses in the opposite direction.
    e) **Gill rakers**—bony structures located on each gill arch and pointing towards the pharynx. Collectively, they act as a strainer preventing food and other particles from being carried across the gills, which could cause the gills injury.

c. **Circulatory system**—the transport system of the body supplies oxygen and nourishment to body cells and removes metabolic wastes

1. Remove the exposed pectoral, pelvis, and anal fins.
2. Using scissors, cut through the lateral side of the body wall.
3. Remove the section of body wall cut.
4. Hold the perch in place by pinning the head, tail, and fins to the dissection tray.

5. Identify the following parts of the circulatory system on the fish and label them on the diagram in Observation 3.

   a) **Heart**—a two-chambered, muscular organ that receives blood from the body, circulates it to the gills for oxygenation to the body, and then back to the heart. Blood in the fish passes from the atrium to the ventricle without being oxygenated.
   b) **Arteries**—carry blood from the heart to the body tissues
   c) **Veins**—carry blood to the heart from the body tissues
   d) **Spleen**—a related circulatory organ; an elongated structure lying on the posterior dorsal surface of the stomach. It is concerned with the production and maintenance of blood cells.

d. **Digestive system**—functions to both mechanically and chemically break down food into simple components that can be easily absorbed by the blood and passed throughout the body to provide body cells with nourishment.

   1. Identify the following parts of the digestive system on the fish and label them on the diagram in Observation 3.

      a) **Esophagus**—a short, straight tube leading from the mouth to the stomach
      b) **Stomach**—receives food from the esophagus and digests the food's protein components
      c) **Intestinal diverticula**—three blind sacs located at the anterior-most part of the duodenum, where it connects to the stomach. The pancreatic and bile ducts open here.
      d) **Duodenum**—partially digested food is passed from the stomach to this organ for final digestion. The duodenum forms an "S"-shaped loop from its origin at the stomach.
      e) **Intestine**—final digestion and absorption of food from the duodenum takes place here
      f) **Fat bodies**—white masses located around the intestine; they function to store food reserves
      g) **Anus**—outside opening of the intestine through which solid waste is passed
      h) **Digestive glands**
         1) **Liver**—reddish-brown, lobed organ situated just anterior and dorsal to the stomach. It produces bile, which aids in fat digestion and the regulation of nutrient levels in the blood.

SECTION 3: NATURAL RESOURCES MANAGEMENT ■ 73

2) **Gallbladder**—greenish sac between the lobes of the liver; stores bile before release to the duodenum.
3) **Pancreas**—a small, folded organ located at the ventral border of the duodenum; manufactures pancreatic juices that further digest proteins

e. **Excretory system**—removes metabolic waste products from the blood.
   1. Identify the following parts of the excretory system on the fish and label them on the diagram in Observation 3.
      a) **Kidneys**—paired, elongated masses pressed against the dorsal body wall
      b) **Collecting ducts**—tubes that drain each kidney. In the male, they connect to the urinary bladder. In the female, they connect the kidney to a urogenital sinus where the urinary bladder is incorporated into the oviduct.
      c) **Urinary papilla**—external projection of the male urinary bladder. Waste products (urine) are eliminated from the body through a urinary pore. In the female, urine is eliminated through the abdominal pore.
      d) **Urinary bladder**—a small sac of the most posterior, ventral area of the body cavity; functions to store urine

f. **Reproductive system**—responsible for the production of sex cells and their delivery to produce offspring; sex organs are separate in the perch.
   1. Identify the following parts of the reproductive system on the fish and label them on the diagram in Observation 3.
      a) **Testes** (male)—paired, lobulated structures, posterior to the duodenum and stomach, and dorsal to the intestine. They produce sperm, which is transported to the outside by a duct called the *vas deferens* and eliminated through the genital pore.
      b) **Ovary** (female)—single, large sac located in the same position in the body as are the testes in the male. The ovary produces white spherical eggs, which fill the cavity to capacity in the spring.
      c) **Oviduct**—a funnel-like structure where the urinary bladder is incorporated. Eggs are expelled through the abdominal pore.
      d) **Other parts**
         1) **Gas bladder** or **swim bladder**—helps regulate depth independently of body motion; located ventral to the kidney and dorsal to the sex organs

74 ■ SECTION 3: NATURAL RESOURCES MANAGEMENT

Observations

1. Label the external parts of the perch on the following diagram.

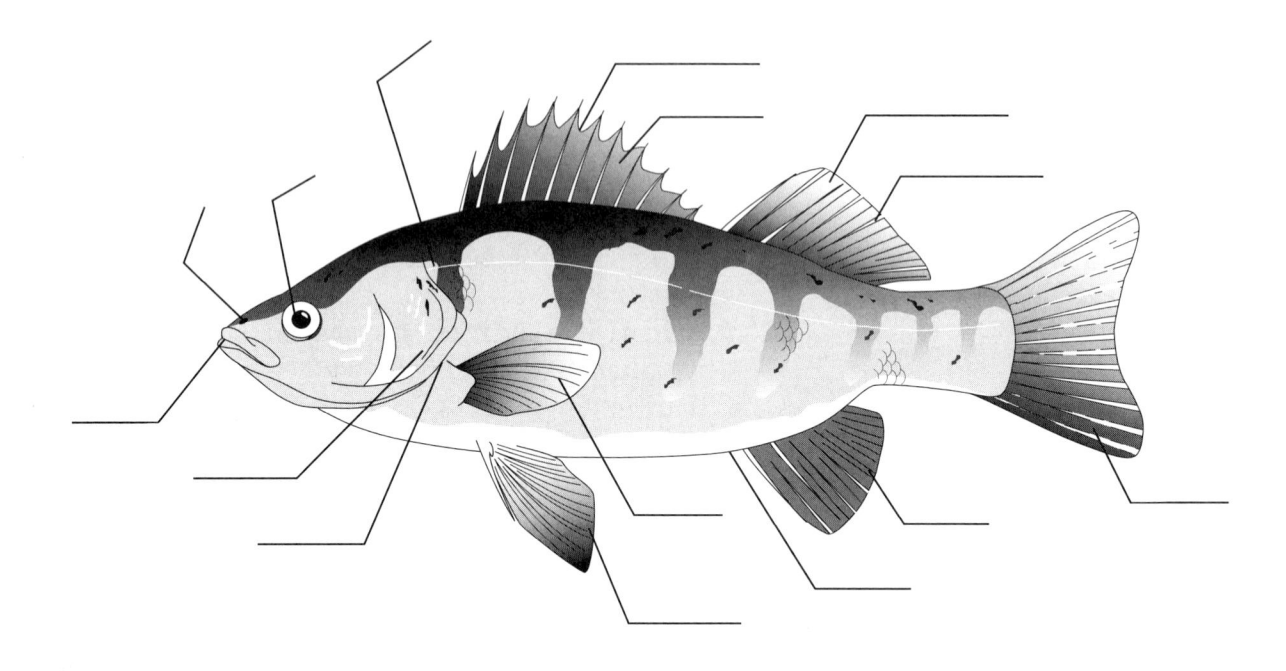

2. Draw a diagram of the fish scale and identify the cteni.

SECTION 3: NATURAL RESOURCES MANAGEMENT ■ 75

3. Label the internal parts of the perch on the following diagram.

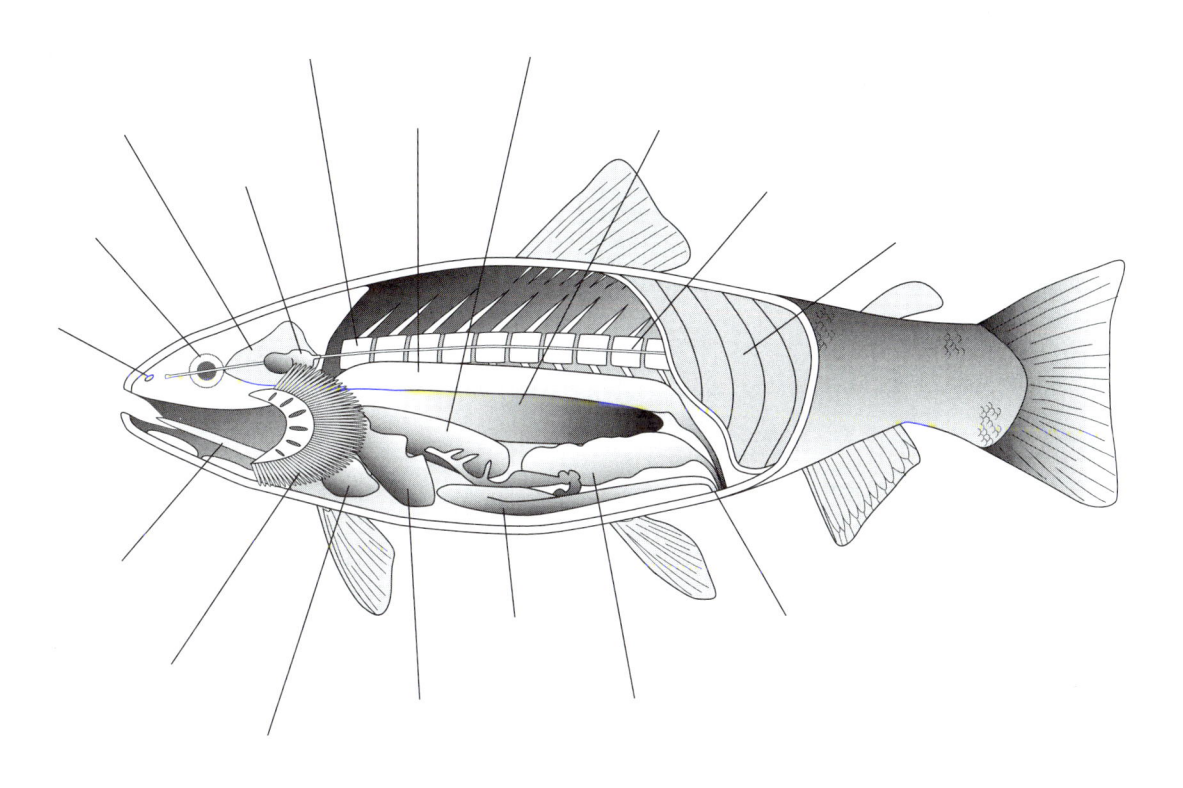

4. Is your fish male or female? How can you tell?
_____
_____
_____

5. Cut open the stomach of the fish. What types of food do you find there?
_____
_____
_____
_____
_____

Conclusions

1. How many fins does your fish have? What purpose is served by the fins?

2. How do the nostrils aid the fish in smelling?

3. Based on the mouth location, what type of feeder is the perch?

4. How is the heart of a fish different from yours?

5. What is the function of the gas bladder?

SECTION: 4

Integrated Pest Management

EXERCISE 17: INSECT ANATOMY

Materials Needed

- Dissection Manual Grasshopper Kit
- One pair of fine dissection scissors
- One sharp probe
- Dissecting microscope
- Personal Safety Set

SAFETY NOTE

Always wear eye protection and rubber gloves when handling dissection specimens. Wash hands before leaving the laboratory.

Purpose To recognize and identify the major external and internal features and structures of the grasshopper and explain their functions.

Procedure

1. Study the information regarding anatomical orientation terminology and the biology of the grasshopper as outlined in the grasshopper kit.

2. Study the information regarding body shape and the exoskeleton as outlined in the grasshopper kit.

3. External anatomy

   a. Place the grasshopper on the dissection tray so that its head is to your left and its legs are down.

   b. Locate and identify the external features of the grasshopper by referring to the diagrams provided in the grasshopper kit.

   c. Use the dissection microscope to observe some of the smaller details.

   d. Locate and label the external parts on the diagram in Observation 1.

4. Identify the sex of your grasshopper using the external anatomical features detailed in the kit. Record in Observation 2.

5. Internal anatomy

   a. Snip off the legs, wings, and antennae of the grasshopper at their bases.

   b. Using pointed iris scissors, insert one point of the scissors on the left side in the intersegmental membrane between the tenth and eleventh abdominal segments just below the cercus. Cut anteriorly up the left side of the grasshopper at a a level of approximately ⅛ inch above the level of the spiracles.

   c. Continue to cut at the same approximate level all the way to the head. (Observe the diagram in the kit.)

   d. Make a similar cut at the same approximate level on the right side of the animal, extending the cut anteriorly from the tenth and eleventh abdominal segment to the head. Join both cuts so that the grasshopper is completely bisected laterally.

e. Carefully lift the posterior part of the dorsal body dissection, working it towards the head, free from the larger ventral body dissection. No difficulties will be encountered until you reach the thorax. Here, cut through the dorsoventral and oblique body muscles and ligaments. In the head, free the head muscles as close to the dorsal body dissection body wall as possible.

f. Remove the dorsal body dissection and place both the dorsal and ventral body dissections in the dissection tray, using insect pins to securely position both dissections.

g. Orient both dissections (ventral and dorsal) in the dissection tray with the head region facing away from you.

h. Locate and identify the internal structures of the grasshopper by referring to the diagrams provided in the grasshopper kit. Study the various systems as discussed in the kit.

i. Label the various internal parts on the diagrams in Observation 3.

Observations

1. Label the external parts of the grasshopper on the following diagram.

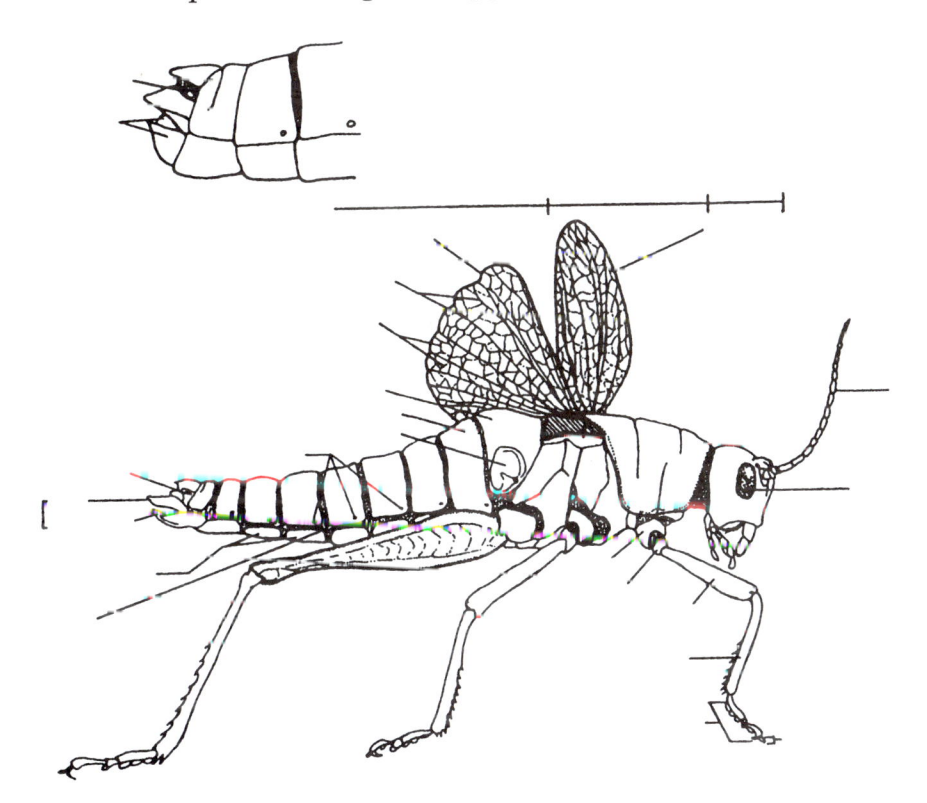

2. Circle the sex of your grasshopper.   Male   Female

SECTION 4: INTEGRATED PEST MANAGEMENT

3. Label the internal parts of the grasshopper on the following diagrams:

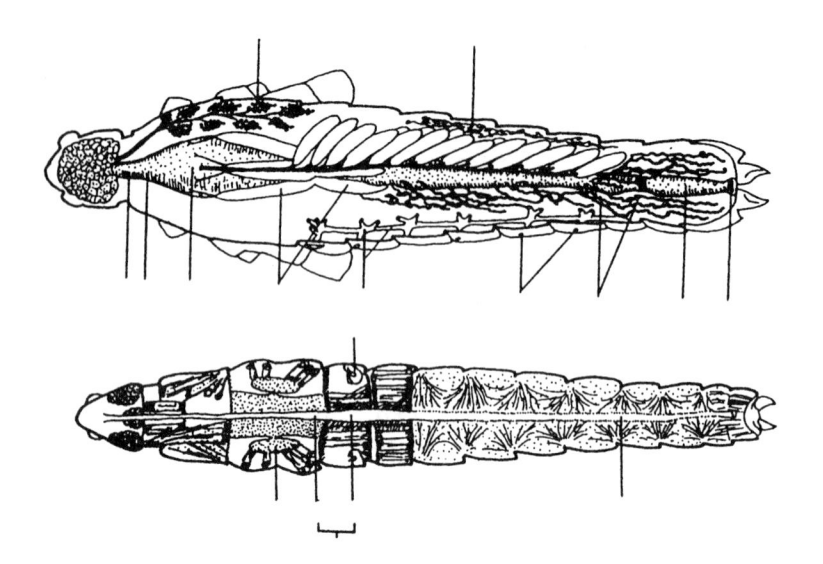

Conclusions

1. How is the exoskeleton of a grasshopper different from the skeleton of a human? How are they alike?

   _____
   _____
   _____

2. What three types of mouthparts may insects have? What type does the grasshopper have?

   _____
   _____
   _____

3. What are the three main body parts of a grasshopper? _____
   _____

4. Explain the functions of the following:
   a. Antenna _____
   b. Mandibles _____
   c. Pretarsus _____
   d. Spiracle _____
   e. Ovipositor _____
   f. Sclerites _____
   g. Crop _____
   h. Malpighian tubules _____
   i. Ocellus _____
   j. Cercus _____

EXERCISE 18:

REGULATING PLANT GROWTH VIA CHEMICALS

Purpose   To observe the effects of various growth hormones on plants.

Procedure

1 Conducting the colchicine experiment.

a. Cut the old roots from the base of the two medium-size onions.

b. Prepare a 0.03% solution of colchicine. Add 33 milliliters of 0.1% solution to 66 milliliters distilled water. Pour into a beaker. Label the beaker "test."

c. Add 100 milliliters distilled water to the other beaker. Label the beaker "control."

d. Using the toothpicks, suspend the onions in each container so that the cut surfaces are just immersed in the liquid.

e. Place both containers in a darkened corner of the room for several days. Check periodically to ensure that liquid remains in contact with the onions. Add a little distilled water if necessary.

f. In several days, new roots will emerge. When they are about 1 inch long, remove them with a razor blade and place one root on a microscope slide.

g. Add several drops of acetocarmine stain to cover the root.

h. Using a pair of forceps, hold the slide directly over an incandescent bulb for 5 minutes. Be careful not to let the stain boil when heating.

i. Cut the very tip from the root and transfer the tip to another slide. Add a fresh drop of acetocarmine stain and mince the root tip with the razor blade.

j. Cover with a coverslip. Press gently on the coverslip to crush the cells. Try not to exert any lateral pressure that will grind or distort the cells.

k. Under the microscope, examine the slides prepared from the treated roots and those prepared from the control. Under high power, you should be able to see the chromosomes of the normal and treated cells.

l. Draw what you see in Observation 1.

Materials Needed

- ✔ Plant Growth Experiment Kit
- ✔ Two medium-size onions
- ✔ Distilled water
- ✔ Two 100-ml beakers
- ✔ Labels
- ✔ Pencil or pen
- ✔ Razor blade
- ✔ Microscope slides
- ✔ Coverslips
- ✔ Nine young tomato plants
- ✔ Toothpicks
- ✔ Forceps
- ✔ Light source
- ✔ Four light-tight boxes
- ✔ Personal Safety Set

SAFETY NOTE

Colchicine is poisonous. Ingestion may be hazardous to your health. Wear protective eyewear and wash hands before leaving the laboratory.

2 Conduct the gibberellic acid experiment.

   a. Label four of the tomato plants "1" through "4."

   b. On plant 1, brush 0.001% gibberellic acid on one leaf of the plant every three to five days for four weeks.

   c. On plant 2, brush 0.001% gibberellic acid on the stem of the plant just above the soil every three to five days for four weeks.

   d. On plant 3, brush 0.001% gibberellic acid on the axial bud of the plant every three to five days for four weeks.

   e. Leave plant 4 to grow normally.

   f. Water the plants as necessary during the experiment.

   g. Record your findings on the chart in Observation 2.

3 Conduct the naphthaleneacetic acid experiment.

   a. Water the plants well so you will not need to disturb them during this experiment.

   b. Arrange the four light-tight boxes so their holes face different directions. For example, one box has its hole toward the east side, one has its hole toward the west side, one has its hole toward the north side, and one has its hole toward the south side.

   c. Place a light source (a lamp will work) in the middle of the four boxes.

   d. Place a tomato plant in each box.

   e. Place one tomato plant where it gets light from all directions.

   f. Treat two tomato plants in opposite boxes by coating their illuminated sides with a lanolin mixture of NAA once daily or an aqueous solution of NAA two or three times daily for 1 week.

   g. Do not treat any of the other plants.

   h. At the end of a week, record your findings on the chart in Observation 3.

SECTION 4: INTEGRATED PEST MANAGEMENT ■ 83

Observations

1. Draw what you see under the microscope.

2. Complete the following chart with your findings.

Plant Number	Week 1	Week 2	Week 3	Week 4
1				
2				
3				
4				

3. Complete the following chart with your findings.

Plant Number	Observations
1	
2	
3	
4	

Conclusions

1. What effect did the colchicine have on the chromosomes in the roots of the onion plant?

2. What effect would this type of change have on the plant?

3. What effect did the gibberellic acid have on the various parts of the plants?

4. How might gibberellic acid be useful to plant growers?

5. What happened to the plants treated with the naphthaleneacetic acid?

6. How did they compare to the untreated plants?

EXERCISE 19:

EXAMINING ECOSYSTEMS

Materials Needed

✔ Study of Ecosystems Kit

✔ Personal Safety Set

SAFETY NOTE

Wash hands before leaving the laboratory.

Purpose To study the relationships of various organisms to each other in an ecosystem.

Procedure

1. Read through both the general aseptic technique handout and the kit instructions.

2. Pour one plate of each carbon source, one plate of each pH level, and one extra pH-7 plate (a total of 7 plates).

3. Rehydrate and dilute the cultures of bacteria using the procedure outlined in the ecosystem kit.

4. Prepare the control plate.

    a. Using a marking pen, label the bottom of one pH-7 plate with a "C" for control.

    b. Squeeze one drop of the microorganisms on the agar edge away from you. Using a sterile spreader, streak the culture back and forth from edge to edge.

    c. Move the spreader toward you in parallel lines to the center of the plate aseptic, per the technique mentioned in Procedure 3; then turn the plate 180 degrees and continue streaking, moving the spreader away from you to the other side. Close the cover immediately.

5. Prepare the experimental carbon source plates.

    a. Label the bottom of each of the special carbon source plates with a code for the carbon, for example, "G" for glucose.

    b. Using the aseptic technique mentioned in Procedure 3, place one drop of the mixed microorganisms onto the agar and spread the drop evenly over the plate. Cover the plate immediately.

6. Prepare the experimental pH plates.

    a. Label the bottom of each of the pH plates with the appropriate pH number (5, 7, or 9).

    b. Using the aseptic technique mentioned in Procedure 3, place one drop of the mixed organisms onto the agar and spread the drop evenly over the plate. Cover the plate immediately.

7. Allow the inoculum to dry on the agar surface for 5 minutes and incubate upside down.

8. Check the plates at 5 and 10 days and record your findings on the chart in Observation 1.

Observations

1. Compare the size, color, and surface appearance (dry, rough, smooth, glistening) on the carbon source plates and the pH plates to those on the control plate. Record your findings in the following chart.

	Control	Carbon Source	pH Level
Color			
Size			
Surface			

2. Use the dichotomous key in the ecosystem kit and the information in the preceding chart to identify the three microorganisms included in the mini-ecosystem. _____

Conclusions

1. What evidence do you have that organisms on the carbon source plates were competing for space and food?

2. Is there one organism for which pH made no difference? If so, which one?

3. Did any one organism grow well on all six experimental plates (three carbon source and three pH levels)?

4. How does this mini-ecosystem compare to a larger animal ecosystem?

SECTION: 5

Plant Sciences

EXERCISE 20:

STEM ANATOMY

Materials Needed
- Monocot stem slide
- Dicot stem slide
- Compound microscope
- Set of colored pencils

Purpose To study the structures and functions of the parts of monocot and dicot stems.

Procedure

1. Study the functions of the monocot stem parts described in the following and identify them on the labeled diagram of a stem cross section.

   a. **Epidermis** — the hard, green, outer layer of a monocot stem; functions to protect the stem from loss of water and from plant infections

   b. **Fibro vascular bundle** — collectively, fibrous strands found in the monocot stem and composed of xylem, phloem, and mechanical tissue

   c. **Xylem** — the inner part of the vascular bundle; carries water and nutrients from the roots to the other parts of the plant

   d. **Phloem** — the small cells on the outer part of the vascular bundle; carry food produced by photosynthesis in the leaf to other parts of the plant. The phloem is composed of sieve tubes, which are elongated, square-cornered cells of varying sizes and with perforated ends.

   e. **Mechanical tissue** — a layer of thick-walled cells surrounding the vascular bundle; provides support and protection

   f. **Parenchyma** — cells that make up the pith of a monocot plant

   g. **Air space** — air spaces found in the pith of the monocot stem

2. Place the monocot stem cross section on the microscope stage centered over the light hole and fasten with the stage clips. Identify the stem parts under the microscope.

3. Label the parts of the monocot stem on the diagram in Observation 1 and color the parts as follows:

   a. Epidermis — light green
   b. Xylem — red
   c. Phloem — yellow
   d. Mechanical tissue — brown
   e. Air space — white
   f. Parenchyma — orange

④ Study the functions of the dicot stem parts described in the following and identify them on the labeled diagram of a stem cross section.

   a. **Cork** — outer layer of the bark; serves to protect the stem of the plant from disease and loss of water

   b. **Cork cambium** — inner layer of the bark; gives rise to cells that become impregnated with a waxy substance, forming a waterproof layer that becomes cork

   c. **Cortex** — cells that are found inside the cork cambium and serve as a food storage area

   d. **Phloem** — the layer of cells that is found inside the cortex and serves as a food transport system. The phloem consists of phloem rays and phloem fiber.

   e. **Cambium** — cylinder of thin-walled cells that gives rise to new xylem to the inside and phloem to the outside

   f. **Xylem** — tissue that carries water and forms the wood of a tree. The growth rings of a tree are found here

   g. **Pith** — the center of a dicot stem; provides support to the plant during the first year of growth

   h. **Vascular rays (phloem rays and xylem rays)** — horizontal strips of thin-walled cells; provide for the transverse conduction of water and nutrients along the tissues of the stem

⑤ Place the dicot stem cross section on the microscope stage centered over the light hole and fasten with the stage clips. Identify the stem parts under the microscope.

⑥ Label the parts of the dicot stem on the diagram in Observation 2 and color the parts as follows:

   a. Cork — yellow

   b. Cork cambium — orange

   c. Cortex — light green

   d. Phloem — dark green

   e. Cambium — red

   f. Xylem — brown

   g. Pith — blue

90 ■ SECTION 5: PLANT SCIENCES

Observations

1. Label the parts of the monocot stem on the following diagram and color the parts as outlined in Procedure 3.

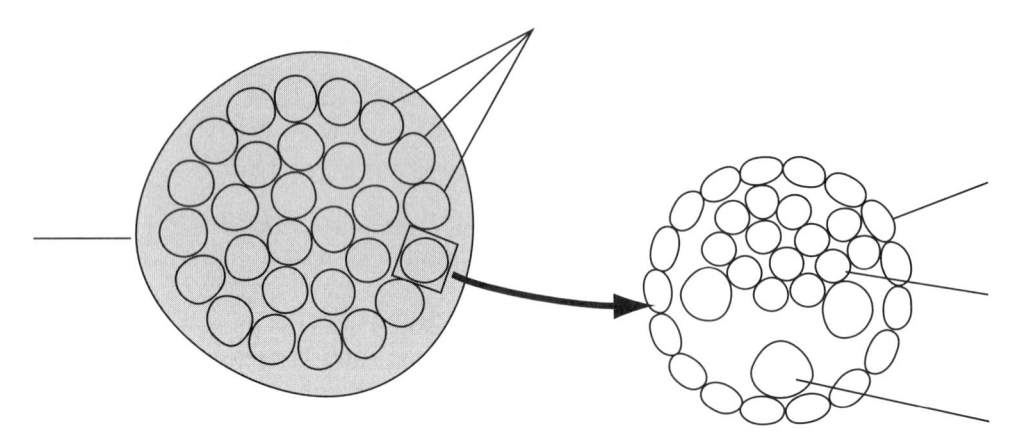

2. Label the parts of the dicot stem on the following diagram and color the parts as outlined in Procedure 6.

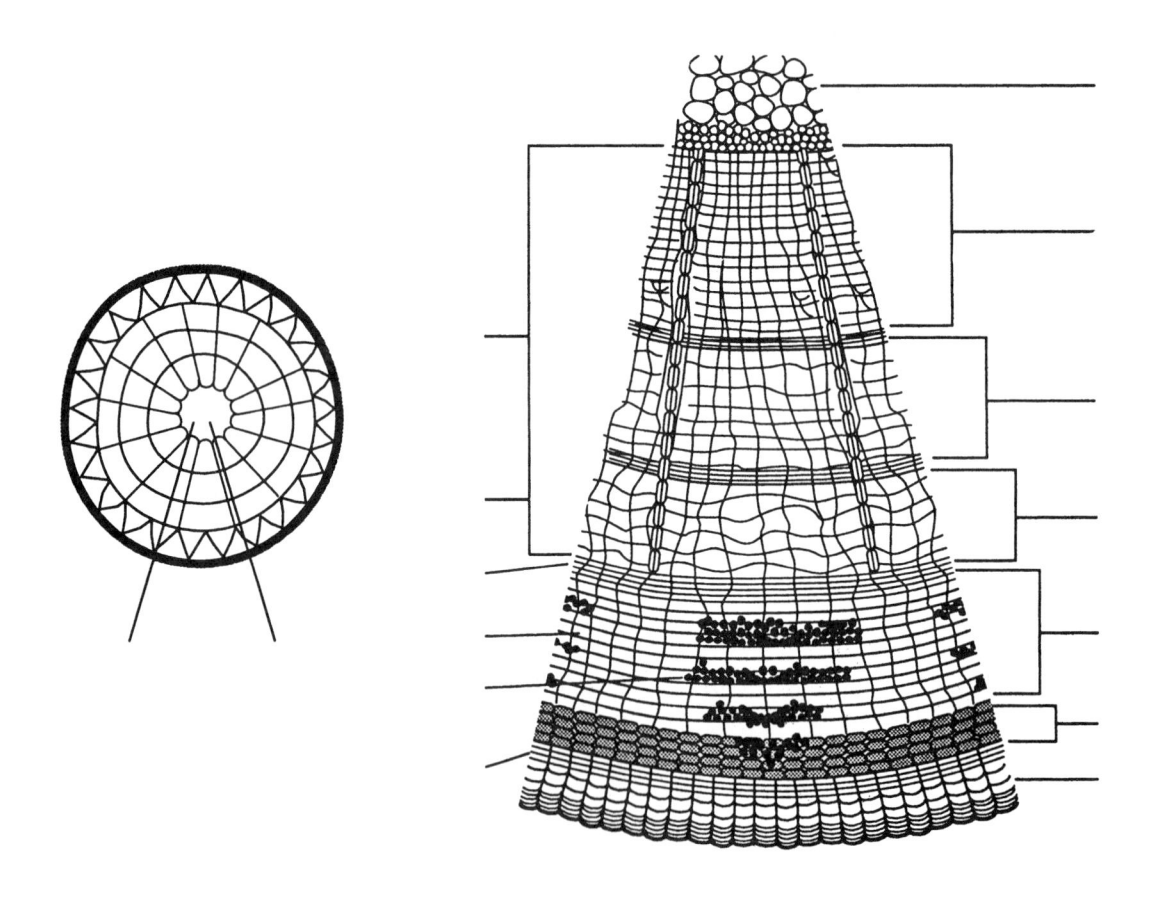

Conclusions

1. What composes most of the cells in a monocot stem? In a dicot stem?

2. What are the functions of the following in a plant stem?

    a. Xylem

    b. Phloem

    c. Cambium

    d. Epidermis

3. How can you determine the age of a dicot? How old is the one that you observed under the microscope?

4. What functions do stems serve?

5. If you girdle a tree (that is, cut a ring around the tree through the bark), why does the tree die?

EXERCISE 21: LEAF ANATOMY

Materials Needed
- Leaf section slide
- Compound microscope
- Set of colored pencils

Purpose To study the structures and functions of parts of a typical plant leaf.

Procedure

1. Study the functions of the leaf parts described following and identify them on the labeled diagram of a leaf cross section.

   a. **Cuticle** — a very thin, noncellular layer of waxy material found on both the top and bottom of the leaf.

   b. **Upper epidermis** — a single layer of heavy, walled cuboidal cells that serve as protection for the top part of the leaf

   c. **Lower epidermis** — a single layer of heavy, walled cuboidal cells that serve as protection for the lower part of the leaf.

   d. **Chloroplasts** — the bodies in a leaf that contain chlorophyll, which imparts the green color to leaves.

   e. **Mesophyll** — the leaf tissue between the upper and lower epidermis; contains palisade cells, air space, and a spongy layer

   f. **Palisade cell** — long, rectangular cells composing a layer one or two cells in thickness just below the upper epidermis; chloroplasts are found in these cells

   g. **Air spaces** — spaces of air found between the spongy layer cells

   h. **Spongy layer** — layer of loosely packed cells that is several cell layers thick; chloroplasts are found in this layer

   i. **Stoma (Stomata)** — openings found in the lower epidermis of a leaf and through which moisture and gases pass

   j. **Guard cells** — crescent-shaped cells that stand out from other epidermal cells because of their thickened walls and because they contain chloroplasts. They control the moisture content of the leaf and the exchange of gases.

   k. **Vascular bundle** — conducts water to the cells of the leaf and carries manufactured food from the leaf to the other parts of the plant. It is really a vein of the leaf itself.

l. **Xylem** — thick-walled, upper-conducting cells in the vascular bundle

m. **Phloem** — thin-walled, lower-conducting cells in the vascular bundle

② Examine a leaf under a microscope.

a. Place a prepared slide of a leaf cross section on the stage. Center it over the light opening and fasten it in place with the stage clips.

b. Focus with the low-power lens and observe the leaf parts. You may want to refer to the diagram in Observation 1.

c. Label the leaf parts on the diagram in Observation 1.

d. Find a vein of the leaf and switch to high power.

e. Observe the xylem and phloem.

③ Color the leaf parts in the diagram in Observation 1 as follows.

a. Cuticle — orange

b. Upper epidermis and lower epidermis — yellow

c. Palisade cells — pink

d. Air space — white

e. Spongy layer — dark green

f. Stoma — light green

g. Guard cells — red

Observations

1. Review the definitions in one of the Procedures and refer to Figure 15–17 in *Agriscience: Fundamentals & Applications, 3E*. Label the parts of the leaf in the following diagram and color the parts as outlined in Procedure 3.

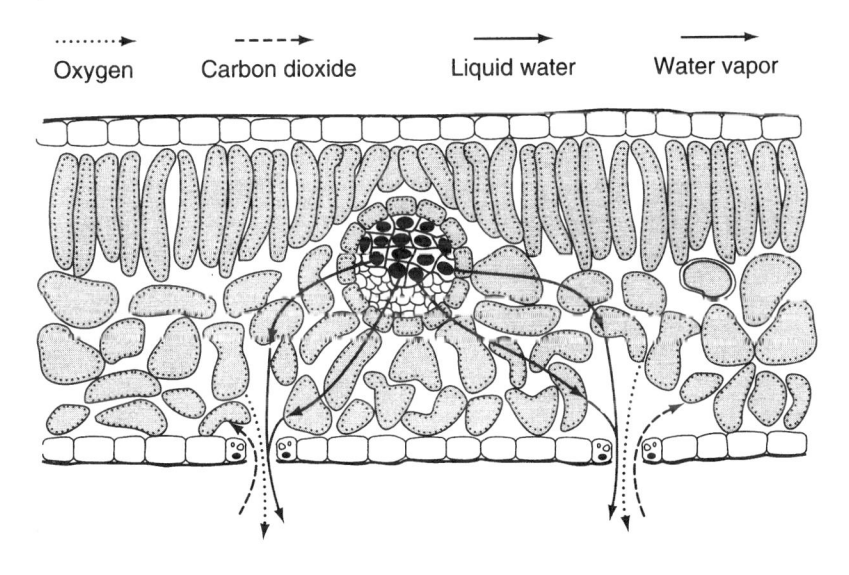

Oxygen   Carbon dioxide   Liquid water   Water vapor

94 ■ SECTION 5: PLANT SCIENCES

Conclusions

1. What is the main function of the leaf in a plant?

2. What function do chloroplasts serve?

3. Can you name some plants whose leaves would not contain chlorophyll?

4. What is the function of each of the following leaf parts?

   a. Epidermis

   b. Stomata

   c. Guard cells

   d. Cuticle

   e. Chloroplasts

EXERCISE 22: EXAMINING PLANT FLOWERS

Materials Needed

- ✔ Taraxacum young bud slide
- ✔ Taraxacum older bud slide
- ✔ Mature anther slide
- ✔ Pollen germination slide
- ✔ Compound microscope
- ✔ Set of colored pencils

Purpose To become familiar with the parts of a plant flower and the functions of the parts.

Procedure

1 Study the following parts and functions while observing the labeled diagram.

  a. **Stamen** — collectively, the male reproductive parts of a plant flower: the filament, anther, and pollen

  b. **Filament** — long stalk that supports the anther. The anther is attached to the end of the filament.

  c. **Anther** — the male sex organ of a plant flower; manufactures the pollen

  d. **Pollen** — the male reproductive sex cell of a plant flower

  e. **Pistil** — collectively, the female reproductive parts of a plant flower: the stigma, style, ovary, and ovules

  f. **Stigma** — the center part of the plant flower; receives the pollen from the anther

  g. **Style** — the pollen enters the stigma and travels down the style to the ovary

  h. **Ovary** — the female sex organ of the plant flower. Ovules are produced here.

  i. **Ovules** — the female reproductive cells of a plant; they are fertilized by the pollen and eventually ripen into seeds

  j. **Perfect flower** — has all of the parts mentioned in 1a through 1i

  k. **Imperfect flower** — at least one but possibly more of the parts mentioned in 1a through 1i are missing

  l. **Petals** — parts of a flower that produce a scent that acts as an attractant to insects and other pollinators

  m. **Sepals** — serve as protection for the developing flower

2 Place the young bud slide under the microscope and view it carefully. Draw the young bud in Observation 1.

96 ■ SECTION 5: PLANT SCIENCES

3 Place the older bud slide under the microscope and view it carefully. Compare it to the young bud slide. Record any differences in Observation 2.

4 Place the mature anther slide under the microscope and view it carefully. Record what you see in Observation 3.

5 View the pollen germination slide. Make a drawing of your observation in Observation 4.

6 Complete the diagram in Observation 5.

Observations

1. Draw the young bud as seen under the microscope.

2. What differences were observed between the young bud and the older bud?
   _____
   _____
   _____
   _____
   _____
   _____

3. Draw the mature anther as seen under the microscope.

4. Draw pollen germination as viewed under the microscope.

5. Label the parts of the following diagram.

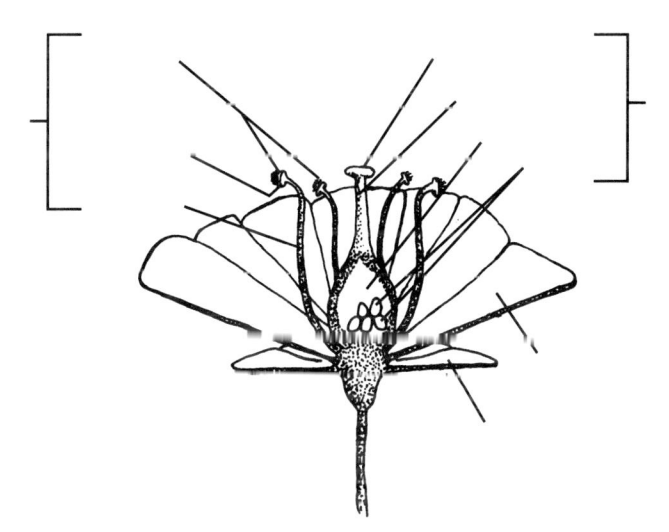

Conclusions

1. Identify the functions of the following flower parts.

    a. Filament _____

    b. Anther _____

    c. Stigma _____

    d. Ovary _____

    e. Petals _____

    f. Sepals _____

2. Explain the difference between a perfect and an imperfect flower. _____
_____
_____
_____
_____
_____
_____

3. Color the parts of the flower diagram in Observation 5 as follows:

    a. Anther — green

    b. Filament — light blue

    c. Pollen — pink

    d. Stigma — purple

    e. Style — brown

    f. Ovary — orange

    g. Ovules — dark blue

    h. Petals — yellow

    i. Sepals — red

EXERCISE 23: GAS PRODUCTION IN PHOTOSYNTHESIS

Materials Needed

- ✓ Calomba plant or other water plant
- ✓ Manometer
- ✓ Water
- ✓ Large beakers
- ✓ High-intensity light source
- ✓ Light-tight box
- ✓ Wax pencil

Purpose To demonstrate the effect of oxygen in photosynthesis.

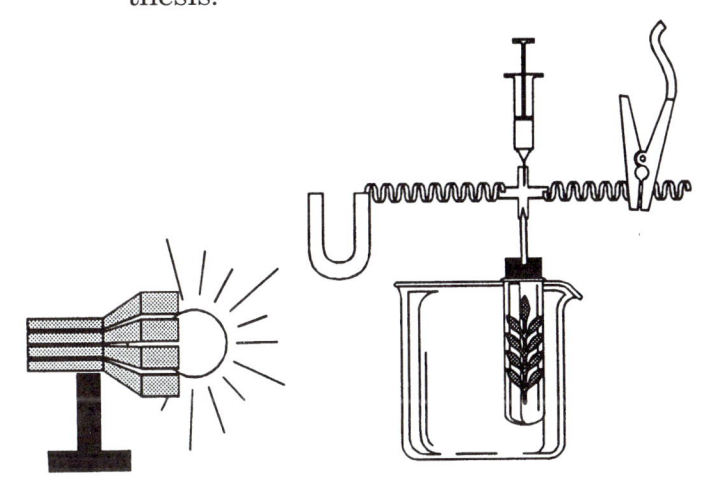

Procedure

1. Place the calomba (or other water plant) in a test tube with water and seal the test tube with a stopper.

2. Complete the oxygen-measuring apparatus (manometer) as shown in the preceding diagram.

3. Place the test tube in the glass container of water, as illustrated in the preceding diagram.

4. Place the light approximately 2 feet from the apparatus on the side farthest away from the test tube.

5. Adjust the syringe as necessary to even the height of the colored fluid in the U-shaped tube to mark the position of the fluid. Use a wax pencil.

6. Let the light shine on the plant for 10 minutes and measure the height of the fluid. Record your findings in Observation 1.

7. Reposition the light 15 feet from the apparatus and use the syringes to readjust the height of the colored fluid in the U-shaped tube until the fluid is again even.

8. Let the light shine on the plant for 10 minutes and measure the height of the fluid. Record your findings in Observation 2.

9. Use the syringe to readjust the height of the level of the colored fluid to even one more time. Place a light-proof box over the apparatus.

10. Wait 10 minutes and measure the height of the fluid. Record your findings in Observation 3.

99

Observations

1. Direction of fluid movement in relation to test tube _____

   Amount of movement (in centimeters) _____

2. Direction of fluid movement in relation to test tube _____

   Amount of movement (in centimeters) _____

3. Direction of fluid movement in relation to test tube _____

   Amount of movement (in centimeters) _____

Conclusions

1. What caused the movement of the fluid in the manometer?

   _____
   _____
   _____
   _____

2. How did the light intensity affect the rate of photosynthesis?

   _____
   _____
   _____
   _____

3. Was any oxygen produced when the plant was in the dark? Why or why not?

   _____
   _____
   _____
   _____

4. What purpose did the beaker of water serve in this experiment?

   _____
   _____
   _____
   _____

EXERCISE 24: WATER MOVEMENT IN PLANTS

Materials Needed

✓ *Osmosis and Diffusion Kit*

Purpose To demonstrate water movement in plants (osmosis and diffusion).

Procedure

① Study the Special Notes section of the osmosis and diffusion kit instructions.

② Fill the plastic container with water to within ¾ inch from the top.

③ Test the water for the presence of glucose as outlined in the osmosis and diffusion kit.

④ Tie a knot very tightly approximately 1 centimeter from the end of the thoroughly moistened and opened membrane tube.

⑤ Put 2.5 to 3 centimeters of glucose solution into the membrane tube.

⑥ Put 2.5 to 3 centimeters of liquid starch into the membrane tube.

⑦ Hold the top of the membrane tube closed by pinching it together and rinse it under running water to remove any glucose or starch on the outside. Gently squeeze the tube to mix the contents.

⑧ Place the filled membrane tube into the container of water, being careful not to spill the contents of the tube, and to keep the open end of the tube outside of the container.

⑨ After 5 minutes, look at the contents inside of the membrane tube and at the liquid in the container. Record your observations in Observation 1.

⑩ Test the water in the container for glucose, as was done in Procedure 3. Record your results in Observation 2.

Observations

1. Record the color of the liquid inside the membrane tube and of the liquid in the container.

Color of Liquid in the Membrane Tube	
Color of Liquid in the Container	

2. Record the results of the glucose test in the following space. Circle the correct response.

   Positive

   Negative

Conclusions

1. What happened to the starch?

2. What happened to the glucose?

3. What materials diffused through the membrane?

4. How was this proven?

5. What substance did not diffuse through the membrane?

EXERCISE 25: ROOT ANATOMY

Materials Needed
- Monocot root slide
- Dicot root slide
- Compound microscope
- Set of colored pencils

Purpose To study the structures and functions of parts of monocot and dicot roots.

Procedure

1. Study the functions of the monocot root parts described following and identify them on the labeled diagram of a root cross section.

   a. **Epidermis** — the outer layer of a monocot root; functions to protect the root from loss of water and from plant infections

   b. **Mechanical tissue (hypodermis)** — a layer of thick-walled cells; serves as support and protection for the root

   c. **Cortical parenchyma** — a thick layer of root cells lying between the hypodermis and the xylem

   d. **Xylem** — composed of tubes of varying size that carry water absorbed by the root hairs from the roots to the other parts of the plant

   e. **Phloem** — the small, closely packed cells found between the xylem and the pith; carry food produced by photosynthesis in the leaf to the roots. The phloem is composed of sieve tubes, which are elongated, square-cornered cells of varying size and with perforated ends.

   f. **Pith** — the region of root cell maturation; composed of undifferentiated parenchyma cells, which are large and not closely packed

   g. **Root tip** — the very short portion of the extreme free end of a root, where growth, water absorption, and the differentiation of the primary tissues take place

   h. **Root cap** — thimble-shaped mass of loose cells that protect the growing point of the root

   i. **Root hairs** — long, fingerlike structures projecting between the tiny particles of soil; absorb water for the entire plant

2. Place the monocot root cross section on the microscope stage centered over the light hole and fasten with the stage clips. Identify the root parts under the microscope.

3 Label the parts of the monocot root on the diagram in Observation 1 and color the parts as follows:

   a. Epidermis — light green
   b. Xylem — red
   c. Phloem — yellow
   d. Mechanical tissue — brown
   e. Cortical parenchyma — orange

4 Study the functions of the dicot root parts described in the following and identify them on the labeled diagram of a root cross section.

   a. **Epidermis** — the outer layer of a dicot root; functions to protect the root from loss of water and from plant infections

   b. **Parenchyma** — a thick layer of root cells lying between the epidermis and the endodermis

   d. **Phloem** — the outer layer of the stele; serves as a food transport system

   e. **Cambium** — cylinder of thin-walled cells that gives rise to new xylem to the inside and phloem to the outside. The second layer of the stele found between the phloem and the xylem

   f. **Xylem** — tissue that carries water from the roots to the other parts of the plant

   g. **Pericycle** — sheath of small, parenchymatous cells that surrounds the xylem and phloem

5 Place the dicot root cross section on the microscope stage centered over the light hole and fasten with the stage clips. Identify the root parts under the microscope.

6 Label the parts of the dicot root on the diagram in Observation 2, and color the parts as follows:

   a. Epidermis — yellow
   b. Parenchyma — orange
   c. Endodermis — light green
   d. Phloem — dark green
   e. Cambium — red
   f. Xylem — brown
   g. Pericycle — blue

SECTION 5: PLANT SCIENCES ■ 105

Observations

1. Label the parts of the monocot root on the following diagram and color the parts as outlined in Procedure 3.

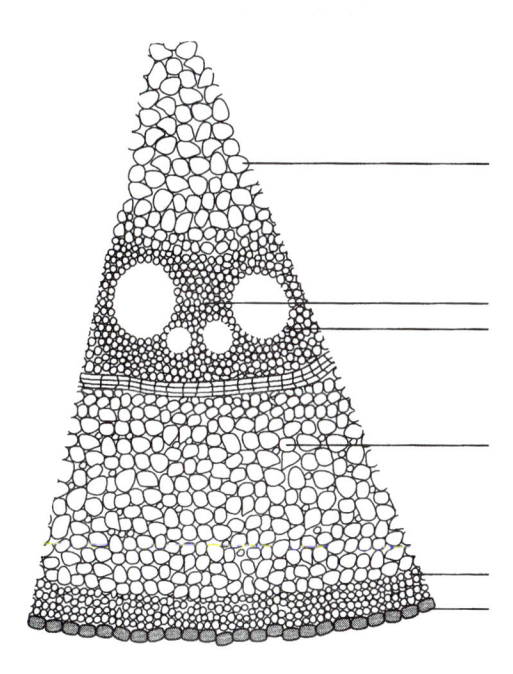

2. Label the parts of the dicot root on the following diagram and color the parts as outlined in Procedure 6.

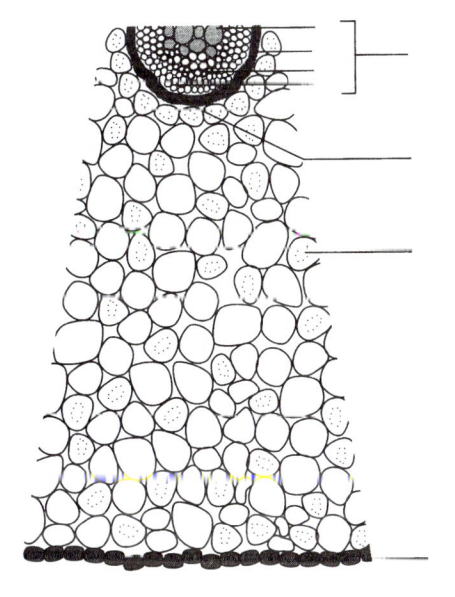

Conclusions

1. What composes most of the cells in a monocot root? In a dicot root?

2. What are the functions of the following in a plant root?

    a. Xylem

    b. Phloem

    c. Cambium

    d. Epidermis

    e. Root cap

    f. Root hairs

3. What functions do roots serve?

EXERCISE 26:

SEED ANATOMY

Materials Needed
- Seed of a plant slide
- Microscope
- Several fresh bean seeds
- Magnifying lens
- Scalpel

Purpose   To become familiar with the parts and functions of the parts of seeds.

Procedure

1. Study the following parts of a seed.
   a. **Fruit coat** — transparent outer layer of the seed
   b. **Endosperm** — makes up most of a seed and is composed of an opaque part, which contains starch, and a transparent part, which is composed of protein. These are the reserve foods that will be used by the embryo when it begins to grow.
   c. **Seed coats** — layers that surround the endosperm and serve to protect the seed's internal parts
   d. **Embryo** — Externally, the part of the seed marked by a slight depression; internally, that part of the seed that grows into a new plant. Composed of the plumule, the plumule sheath, the cotyledon, the radicle, and the radicle sheath.
   e. **Cotyledon** — makes up most of the embryo. The cotyledon's infolded edges almost enclose the plumule. Collectively, the cotyledons are the seed leaves.
   f. **Plumule** — the primary bud of the seed embryo
   g. **Plumule sheath** — made up of several immature leaves, which will become the first leaves of the plant
   h. **Radicle** — becomes the root system of the plant
   i. **Radicle sheath** — sheath of protection that surrounds the radicle
   j. **Point of attachment** — where the seed is attached to the mature plant

2. Place the slide of the corn seed under the microscope and identify the parts. Label these parts on the diagram in Observation 1.

3. Use the scalpel to cut one bean seed as shown in the following diagram. Record what you find inside in Observation 2.

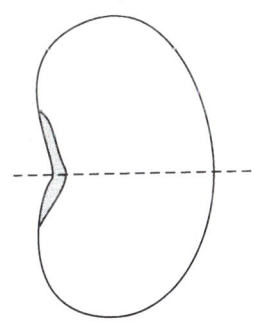

4 Cut another bean seed as shown in the following diagram. Record what you see in Observation 3.

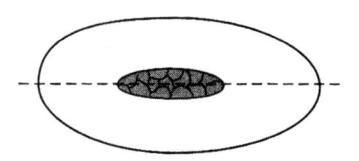

Observations

1. Label the parts of the seed in the following diagram.

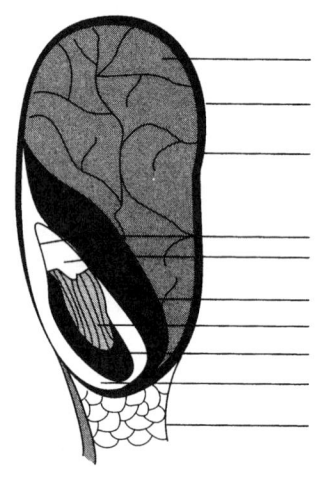

2. Draw what you observed inside the bean seed cut in Procedure 3.

3. Draw what you observed inside the bean seed cut in Procedure 4.

Conclusions

1. What is the function of the largest part of a seed?

2. What are the functions of the following seed parts?

    a. Plumule

    b. Radicle

    c. Embryo

    d. Seed coat

3. What is the function of seeds in the life cycle of a plant?

EXERCISE 27: REQUIREMENTS FOR SEED GERMINATION

Materials Needed
- Water
- Aluminum foil
- Beaker
- Freezer
- Seed-Gro Kit

Purpose   To determine how differences in environmental conditions affect seed germination.

Procedure

1 Moisture experiment

  a. Place a piece of dry aluminum foil in a dry place.

  b. Place five seeds on the foil or paper. *Do not water*.

  c. Place one of the foam germination pads in the petri dish and water as directed by your instructor.

  d. Place three seeds of the same kind as you placed in the foil on the foam pad and add a few drops of water directly on the seeds to hasten germination.

  e. Replace the cover on the germination chamber to reduce moisture loss and keep out unwanted materials.

  f. Observe the seeds for a seven-day period. Record your observations on the chart in Observation 1.

2 Air (oxygen) experiment

  a. Fill a beaker approximately half full with water.

  b. Place five seeds in the beaker.

  c. Place one of the foam germination pads in the petri dish and water as directed by your instructor.

  d. Place three seeds of the same kind as you placed in the beaker on the foam pad and add a few drops of water directly on the seeds to hasten germination.

  e. Replace the cover on the germination chamber to reduce moisture loss and keep out unwanted materials.

  f. Observe the seeds for a seven-day period. Record your observations on the chart in Observation 2.

3 Temperature experiments

  a. Prepare two packets each containing five seeds wrapped in aluminum foil.

  b. Place one packet in the freezer for one week.

  c. Place the other packet in boiling water for 10 minutes.

d. Prepare three germination chambers by placing the foam germination pads in the petri dishes and watering as directed by your instructor.

e. Label the germination chambers "hot," "cold," and "normal."

f. Plant the seeds boiled in water in the chamber labeled "hot," the freezer-conditioned seeds in the chamber labeled "cold," and the five untreated seeds in the chamber labeled "normal."

g. Observe the seeds for a seven-day period.

h. Record your observations on the chart in Observation 3.

4 Light experiment

a. Prepare two germination chambers by placing the foam germination pads in the petri dishes and watering as directed by your instructor.

b. Plant three seeds of the same kind in each chamber.

c. Wrap one chamber in aluminum foil, making sure no light can get into the chamber.

d. Observe the seeds every day for one week. Try to let no light (or as little as possible) get to the seeds in the aluminum-foil-wrapped chamber when you observe them.

e. Record your findings on the chart in Observation 4.

Observations

1. Record your moisture observations in the following chart:

	Day 1	Day 2	Day 3	Day 4	Day 5	Day 6	Day 7
Germination Chamber Seeds							
Foil-Wrapped Seeds							

SECTION 5: PLANT SCIENCES

2. Record the results of the air experiment in the following chart:

	Day 1	Day 2	Day 3	Day 4	Day 5	Day 6	Day 7
Germination Chamber Seeds							
Seeds in Beaker of Water							

3. Record the results of the temperature experiment in the following chart:

	Day 1	Day 2	Day 3	Day 4	Day 5	Day 6	Day 7
Normal Seeds							
Cold Seeds							
Cold Seeds							

4. Record the results of the light experiment in the following chart:

	Day 1	Day 2	Day 3	Day 4	Day 5	Day 6	Day 7
Germination Chamber Seeds							
Foil-Wrapped Seeds							

Conclusions

1. What part does moisture play in seed germination?

2. What happened to the seeds that were placed in the beaker of water? What caused this?

3. What effect did heat have on the germination of the seeds?

4. What effect did cold temperature have on the germination of the seeds?

5. How did the absence of light affect the germination of the seeds?

6. What conditions are necessary for proper seed germination?

EXERCISE 28:

PLANT PROPAGATION

Materials Needed

- ✔ Plant Propagation Kit
- ✔ Peperomia plant
- ✔ Wandering Jew plant
- ✔ Spider or airplane plant

Purpose To observe the growing of plants by the asexual method of plant propagation.

Procedure

1. Approximately 24 hours before you begin to take the first cuttings, prepare materials by opening a bag of growth medium and placing it in the bottom half of the propagator. Add approximately 1 liter of water to the bag and allow this to stand until shortly before use.

2. With the bottom half of the propagator in a sink or tray, empty the bag of wet growth medium into it. Allow the excess water to run out through the drainage holes.

3. Take cuttings as close to sunrise as possible from plants that have been thoroughly watered within the last 24 hours.

4. Do not allow the cuttings to remain in direct sunlight. Instead, store them in any kind of closed container, out of direct sunlight, and with a lining of wet newspapers.

5. Shorten cuttings to 3 to 4 inches and remove leaves from the lower 2 inches of the stems.

6. Using a nail, pencil, or similar instrument, make holes for inserting the cuttings into the growth medium. Firm up the growth medium around the cuttings. *Do not press* the growth medium down.

7. Place as many cuttings as possible in the propagator without creating leaf overlapping.

8. Place the propagator in a window or under fluorescent lights with white reflectors approximately 1 foot over the top of the propagator.

9. Water the cuttings heavily twice each week. If the opaquing shield has been used, it should be removed during the watering process.

10. Observe the cuttings for a period of five to six weeks. Record your findings on the chart in Observation 1.

11. Remove cuttings for transplanting by inserting a flat stick or spoon handle approximately 1 inch from the stem and deep enough so that it will lift the stem and adjacent medium with it. *Do not let cuttings dry out.*

12. Make sure cuttings are potted, well watered, and kept shaded for several days.

Observations

1. Record your observations of the cuttings in the following chart.

	Plants		
	Peperomia	Wandering Jew	Spider or Airplane
Week 1			
Week 2			
Week 3			
Week 4			
Week 5			
Week 6			

Conclusions

1. Which method of plant reproduction is propagation?

_____
_____
_____
_____
_____
_____
_____
_____

SECTION 5: PLANT SCIENCES

2. Can all plants be reproduced by propagation? Why or why not?

3. What happened to the cuttings that were placed in the growing medium?

4. Which kind of plant grew the fastest?

SECTION: 6

Crop Science

EXERCISE 29: SOIL ORGANISMS AND HUMUS

Materials Needed

- ✔ What's in the Soil II— Microorganism Kit
- ✔ Compound microscope—300X or better
- ✔ Four 400-ml beakers
- ✔ Test tube rack
- ✔ Two test tubes
- ✔ Bunsen burner
- ✔ Handful of green leaves
- ✔ Soil samples
- ✔ Water
- ✔ Personal Safety Set

SAFETY NOTE

Always wear eye protection and rubber gloves when handling specimens. Wash hands before leaving the laboratory.

Purpose To recognize the different organisms found in the soil that aid in decomposition of organic matter and the formation of humus.

Procedure

1. Study the information on soil organisms provided in the microorganism kit.

2. Collect a soil sample to be used in this experiment. Record the information regarding the collection site in the table in Observation 1.

3. Label one test tube "broth." Add nutrient broth to the test tube, filling the tube approximately two-thirds full. Set the tube aside.

4. Label one test tube "agar." Pour the agar into the test tube and set the tube at a slant to allow the agar to harden. (Hardening at a slant provides a greater surface for organism growth.)

5. Place soil from your soil sample into a cup and add a small amount of water. Swirl the cup gently to mix the soil and water. Add more water so that the cup is two-thirds full.

6. Use a clean eye dropper and add a drop of the soil-water mixture to the tube containing the nutrient broth. Seal the tube and swirl it gently to mix the contents. Set the test tube in the rack. Check the tube in 24 hours. Record your results in Observation 2.

7. Sprinkle a small amount of soil on the agar in the slanted test tube. Seal the tube and place the test tube in the rack. Check the test tube in 48 hours. Record the results in Observation 3.

8. Wash a small piece of green leaf in water. Rinse the leaf in alcohol to remove any organisms from its surface and rinse again in water.

    WARNING: Alcohol is flammable. Do not expose to heat or flame.

9. Add the leaf to the cup containing the soil water and place the cup in a location that receives sunlight. Be sure that you have labeled the cup with your name. Check the cup in 48 hours. Record your findings in Observation 4.

10 Place a drop of water from the nutrient broth mixture on a glass slide, add a drop of methylene blue, and place a coverslip over the slide. Place the slide on the microscope and view first under low power and then under high power. Record your findings in Observation 5.

11 Place a piece of the growth from the agar solution on a slide, add a drop of methylene blue, and place a coverslip over the slide. Place the slide on the microscope and view first under low power and then under high power. Record your findings in Observation 6.

12 With an eye dropper, remove a small amount of water from near the surface of the leaf in the cup of soil. Place a drop on a slide and cover with a coverslip. View the slide under the microscope and record your findings in Observation 7.

Observations

1. Record the following information regarding the soil sample site:
   a. Location of sample _____
   b. Light conditions _____
   c. Temperature conditions _____
   d. Humidity conditions _____

2. Record your findings regarding the nutrient broth.
   _____
   _____
   _____

3. Record your findings regarding the agar test tube.
   _____
   _____
   _____

4. Record your observations regarding the soil water.
   _____
   _____
   _____

5. Draw or describe what you see on the slide of the nutrient broth and methylene blue.

6. Draw or describe what you see on the slide of the agar and methylene blue.

7. Draw or describe what you see on the slide of the soil water.

Conclusions

1. What soil organisms were observed in this experiment?

2. Why are these organisms found near the soil surface?

3. What functions do these organisms serve in the soil?

4. Were any of the organisms that you found plants? How can you tell?

EXERCISE 30: ENRICHING SOIL THROUGH DECOMPOSITION

Materials Needed
- ✔ Decomposition Kit
- ✔ Flower pots
- ✔ Plastic trays

Purpose — To observe how decomposition adds enrichment to the soil.

Procedure

1. Place a small amount of gravel in the base of four flower pots.

2. Fill two flower pots each with sand up to ½ inch from the top, packing the sand tightly.

3. Fill two flower pots each with soil rich in organic matter.

4. Label the flower pots "1" through "4" and add the following to the pots:

    a. pot 1 (with sand) — add dried plant material

    b. pot 2 (with sand) — add animal material

    c. pot 3 (with organic matter) — add dried plant material. Do not cover more than 50 percent of the surface with organic matter.

    d. pot 4 (with organic matter) — add animal material. Again do not cover more than 50 percent of the surface with organic matter.

    NOTE: Other materials in the kit may also be put in pots.

5. Cover each of the pots with an upside down plastic cover dish, pressing lightly into the sand or soil.

6. Place the pots in a plastic tray. Add water to the tray to a depth above the first lip. Maintain this depth throughout the entire experiment.

7. Place the tray with the pots in a dark area that is at room temperature.

8. Observe the flower pots once a week for a period of eight weeks.

9. Record your findings in the table in Observation 1.

Observations

1. Complete the following table with your findings.

Week Number	Pot Number			
	1	2	3	4
1				
2				
3				
4				
5				
6				
7				
8				

Conclusions

1. Did the animal materials or plant materials decompose faster?

2. Compare the decomposition rate in organic matter to that in sand.

3. Which soil does a better job of aiding in decomposition? Why?

4. What does organic matter do for the soil?

EXERCISE 31: GROWING AND TRANSPLANTING VEGETABLE SEEDLINGS

Materials Needed

- ✔ Tomato seeds
- ✔ Seed starting media
- ✔ Flats
- ✔ Spray bottle with water
- ✔ Trowel or spoon
- ✔ Plastic wrap
- ✔ Large containers (10 ounces each or larger)
- ✔ Plant growing media
- ✔ Liquid fertilizer
- ✔ Ruler

Purpose   To practice the proper methods of sprouting and transplanting seedlings

Procedure

1 Study the following terms.

   a. **Flat** — a wooden or plastic box with slotted bottom used to start seedlings

   b. **Germinate** — to sprout or begin to grow

   c. **Media** — a material which is used to start and grow seeds and plants

   d. **Seedlings** — young plants which have been germinated several days

   e. **Transplant** — to move plants from one growing location to another thus giving them more space in which to develop

2 Growing the seedlings

   a. Fill the sections of the flat with the seed starting media almost to the top.

   b. Place something under the flat to catch any excess water.

   c. Use the spray bottle to mist the mixture until it is thoroughly damp.

   d. Place two tomato seeds in each compartment of the flat or in your assigned compartments if sharing a flat with a classmate.

   e. Use the trowel or spoon and sprinkle additional seed-starting media on top of the seeds no more than ⅛-inch deep.

   f. Mist the mixture again.

   g. Wait 10 to 15 minutes and mist once more with the spray bottle.

   h. Cover the flat with plastic wrap. Do not cover too tightly.

   i. Place the flat in a nice warm spot. Check daily for moisture. Mist if necessary.

   j. Once they sprout, they need light for growth, either artificial or sunlight.

   k. Once the tomatoes have grown their first set of true leaves, they are ready to be transplanted to larger containers.

3 Transplanting the seedlings

   a. Place about 8 ounces of growing media in each container. (Containers should have drainage holes.)

   b. Use the trowel or spoon to gently pry up the tomato seedlings.

   c. Gently hold the seedling by its leaves (not stems) and carefully place it in its new container.

   d. Firm the soil around it with your fingers.

   e. Use the spray bottle to moisten the soil in the container thoroughly.

   f. Make sure the containers are set where they receive adequate light and keep the soil moist (not wet).

   g. After a week, provide the plants with a diluted (one-fourth to one-half strength) mixture of liquid fertilizer.

   h. Plants may be transplanted to their final growing location in about one month.

Observations

1. Record the progress of your seed growth in the chart that follows. Record height and number.

Tomato Seeds' Progress	Day 2	Day 4	Day 6	Day 8	Day 10	Day 12
Number Sprouting						
Height of Tallest						
Height of Shortest						

2. Record the progress of your transplanted tomato plants for three weeks.

Tomato Plants' Progress	Day 2	Day 4	Day 6	Day 7	Day 14	Day 21
Height of Tallest						
Height of Shortest						

Conclusions

1. What were the important environmental conditions necessary for the sprouting of the seedlings?

2. What were the important environmental conditions necessary for the growth of the seedlings after sprouting?

3. What was your germination rate? Hint: Number of seedlings/Number of seeds planted × 100.

4. If the germination rate was not 100 percent, give possible reasons.

5. Did all plants that were transplanted survive? Why or why not?

EXERCISE 32: GRAFTING FRUIT TREES

Materials Needed

- ✓ Knife
- ✓ Grafting wax
- ✓ Grafting tape
- ✓ Apple or other fruit stock—¼-inch to ½-inch diameter
- ✓ Scions — ¼-inch to ½-inch diameter

Purpose To become familiar with the whip and tongue method of grafting fruit trees.

Procedure

1 Study the following terms related to grafting.

a. **Grafting** — uniting two different plants so that they grow as one

b. **Cambium** — thin, green, actively growing tissue located between the bark and wood of a plant

c. **Grafting wax** — a pliable, sticky, waterproof material made of beeswax, resin, and tallow

d. **Rootstock** — the lower portion of a graft that becomes the stem and roots of the new plant

e. **Scion** — a short piece of shoot containing several buds that becomes the new top of a grafted plant.

2 Making the cuts

a. Be sure that the rootstock and scion are as close as possible to the same size.

b. Using the knife, make the first cut on the scion below a bud. The length of the cut should be such that a smooth slanted angle is cut from 1½ to 2 inches long.

c. The rootstock should be cut next with the same length and slope as that of the scion so that the two parts will fit together evenly.

d. On the scion make a second cut beginning about one-third of the distance from the tip. The cut should begin vertically but gradually become nearly parallel to the first cut surface. This is the tongue.

e. Follow the same procedure as in the previous step with the rootstock. When finished, the scion and rootstock should fit together with the tongues interlocking.

3 Completing the graft

a. Make sure that the cambium layers of the rootstock and scion are matched together on at least one side.

b. Push the scion and rootstock together tightly matching the surfaces as closely as possible. Be sure there is no overlap of scion and rootstock.

c. Wrap the graft with the grafting tape to keep it tight to prevent drying.

d. Apply grafting wax over the entire area that is wrapped with tape as uniformly as possible. This is an additional precaution to prevent drying.

e. In both the taping and waxing steps be careful not to dislodge the aligned cambion layers of the scion and rootstock.

Observations

1. In grafting the scion and rootstock together, why was it important to match the cambium layers together?

2. What could be the possible consequences of not wrapping and waxing the graft?

Conclusions

1. Define the following terms:
   a. Cambium

   b. Grafting

   c. Rootstock

   d. Scion

2. Why is grafting used in the fruit and nut industries?

3. Is grafting an economical method of producing fruits and nuts? Explain your answer.

EXERCISE 33:

EFFECTS OF SEED PLANTING DEPTHS ON CROP PRODUCTION

Materials Needed

- ✔ Six glass jars— Pint size
- ✔ Soil
- ✔ Water
- ✔ Corn seeds

Purpose To observe the effects of planting depth on seed germination and crop production.

Procedure

1 Prepare the jars.

a. Label the jars "1" through "6." Prepare the jars as follows:

1. jar 1 — place 1 inch of loosely packed soil in the bottom of the jar and place one corn seed on each of the four sides of the jar next to the side where they can be seen; put 3 inches of loosely packed soil on top of the corn seeds.

2. jar 2 — place 2 inches of loosely packed soil in the bottom of the jar and place one corn seed on each of the four sides of the jar next to the side where they can be seen; put 2 inches of loosely packed soil on top of the corn seeds.

3. jar 3 — place 2½ inches of loosely packed soil in the bottom of the jar and place one corn seed on each of the four sides of the jar next to the side where they can be seen; put 1½ inches of loosely packed soil on top of the corn seeds.

4. jar 4 — place 3 inches of loosely packed soil in the bottom of the jar and place one corn seed on each of the four sides of the jar next to the side where they can be seen; put 1 inch of loosely packed soil on top of the corn seeds.

5. jar 5 — place 3½ inches of loosely packed soil in the bottom of the jar and place one corn seed on each of the four sides of the jar next to the side where they can be seen; put ½ inch of loosely packed soil on top of the corn seeds.

6. jar 6 — place 4 inches of loosely packed soil in the bottom of the jar and place one corn seed on each of the four sides of the jar next to the side where they can be seen; do not cover.

2 Moisten the soil thoroughly but not excessively.

3 Place the six jars where they will receive adequate light and warmth.

SECTION 6: CROP SCIENCE ■ 131

4 Monitor the progress of the seeds for three weeks, watering when needed.

5 Record your findings on the chart in Observation 1.

Observations

1. Record the progress of your seeds in the following chart.

Jar Number	Week 1	Week 2	Week 3
1			
2			
3			
4			
5			
6			

Conclusions

1. Which planting depth (or depths) was/were most appropriate for the corn seed? Why?

___
___
___
___
___

2. What happened to the seeds that were planted the deepest?

___
___
___
___
___

132 ■ SECTION 6: CROP SCIENCE

3. Why is it important to crop production for seeds to be planted at the proper depth?

4. If planted in a field, what would probably have happened to the seeds in jar 6?

5. What conditions are necessary for proper seed germination?

EXERCISE 34: STORING FORAGE AS SILAGE

Materials Needed
- ✔ Freshly cut chopped grass
- ✔ Six 1-quart jars with tight-fitting lids
- ✔ Labels for the jars
- ✔ Pen
- ✔ Nail
- ✔ Hammer

Purpose To observe how silage is produced and how different factors affect silage quality.

Procedure

1 Preparing the forage

a. Spread the grass out to dry for three days.

b. Label the Jars "1" through "6."

c. Ensile the grass as follows:
   1. jar 1 — pack the grass tightly, fill with water, and cover the jar tightly
   2. jar 2 — pack the grass tightly, fill with water, cover the jar, and punch several holes in the lid
   3. jar 3 — pack the grass tightly, cover the jar tightly
   4. jar 4 — pack the grass tightly, cover the jar, and punch several holes in the lid
   5. jar 5 — pack the grass loosely, cover the jar
   6. jar 6 — pack the grass loosely, do not cover the jar

2 Storing the silage

a. Put the jars in a dark place

b. Check them every day for three weeks to observe the development of the silage.

c. Compare the silage in your jars to the quality factors listed in the following table.

d. Record your results in Observation 1.

Comparing Silage Color and Odor			
Color		*Odor*	
Natural green color. Some clovers may turn a darker color than normal	Good	Clean, pleasant, vinegary or pickled smell	Good
Light green or yellowish color. Also may be slightly brown	Fair	Slightly fruity, yeasty, or scorched smell. Corn with high moisture may have acidic smell.	Fair
Gray or whitish color (molding). Black or dark brown (overheating)	Poor	Musty smell (molding). Strong burnt odor (overheating). Rotten odor (poor bacterial action)	Poor

Observations

1. Complete the following chart regarding the quality characteristics of the silage in your jars.

Jar Number	Color	Odor	Other Observations
1			
2			
3			
4			
5			
6			

Conclusions

1. Which jar produced the best silage based on color and odor?

2. What happens when forage is ensiled too wet?

3. Why do you think silage is packed tightly when stored?

SECTION: 7

Ornamental Use of Plants

EXERCISE 35: RESPONSES OF PLANTS TO LIGHT

Materials Needed
- Six Pots of pea seedlings
- Four tight boxes, each with a hole in one side
- One completely light-tight box
- Light source

Purpose   To determine how plants respond to different lighting conditions.

Procedure

1. Water the plants well so you will not need to disturb them during this experiment.

2. Arrange the four tight boxes so their holes face different directions. For example, one box has its hole toward the east side, one has its hole toward the west side, one has its hole toward the north side, and one has its hole toward the south side.

3. Place a light source (a lamp will work) in the middle of the four boxes.

4. Place a pot of pea seedlings in each box.

5. Place one pot of pea seedlings where it gets light from all directions.

6. Place one pot of pea seedlings in the completely light-tight box.

7. Leave the plants in the boxes for one week. At the end of the week, open the boxes and observe the plants.

8. Record the plants' responses to light on the chart in Observation 1.

Observations

1. Complete the following chart regarding the plants' response to light.

Lighting Condition	Plant Response		
	Color of Plant	Direction of Growth	Length of Stem
Light from North			
Light from South			
Light from West			
Light from All Directions			
No Light			

Conclusions

1. What was the general effect of light versus darkness on the pea plants' growth? How can you explain this?

2. How did the color vary on the plants exposed to light versus the one kept in total darkness?

3. How did the plants move in response to the light?

EXERCISE 36:

CLONING PLANTS FOR UNIFORMITY

Materials Needed

- Plant II Cloning: Propagation of African Violets Kit
- African violet plant
- Soap
- Bunsen burner
- Scalpel
- Water—Sterile
- Four to five large beakers
- African violet potting soil
- Ethyl alcohol
- Hot plate
- Laundry bleach
- Ten to twelve small plastic pots
- Personal Safety Set

SAFETY NOTE

Always wear eye protection and rubber gloves when handling dissection specimens. Wash hands before leaving the laboratory.

Purpose To observe the methods of cloning plants to obtain uniformity in production.

Procedure

1. Study the sterile-technique procedures outlined in the cloning kit instruction guide. Follow these procedures when working with this equipment.

2. Follow the procedure outlined in the kit for preparation of shoot multiplication media.

3. Prepare the leaves for culture as described in the kit instruction guide.

4. Establish the shoot multiplication culture as outlined in the guide. Observe the plants every two weeks for three months. Record your findings on the chart in Observation 1.

5. Prepare the rooting medium according to the directions in the kit.

6. Transfer the plants to the rooting medium as outlined in the instruction guide found in the kit. Observe the plants for the next three weeks. Record your findings in Observation 2.

7. After numerous roots have appeared, carefully transfer the plants to pots containing African violet potting soil. Scrape all excess agar from the roots. Treat the young plants as you would any other plant.

Observations

1. Record the plant progress in the following chart.

First 2 Weeks	Second 2 Weeks	Third 2 Weeks	Fourth 2 Weeks	Fifth 2 Weeks	Sixth 2 Weeks

2. Record your findings in the following chart.

First Week	Second Week	Third Week

Conclusions

1. What type of plant reproduction is cloning?

SECTION 7: ORNAMENTAL USE OF PLANTS

2. What advantages can you think of for cloning plants? What disadvantages?

3. Do you think cloning will work for animals?

4. Do you know of any examples of animal cloning? Discuss.

SECTION: 8

Animal Sciences

EXERCISE 37: SIMPLE DIGESTION IN ANIMALS

Materials Needed

- ✔ Digestive tract of pig or fetal pig
- ✔ Rubber gloves
- ✔ Scalpel
- ✔ Dissecting microscope
- ✔ Personal Safety Set

SAFETY NOTE

Always wear eye protection and rubber gloves when handling dissection specimens. Wash hands before leaving laboratory.

Purpose   To identify the parts and the functions of the parts of the simple digestive system.

Procedure

1 Study the following parts of the simple digestive system in Figure 26–9 in *Agriscience: Fundamentals & Applications, 3E* and locate them on the drawing.

Alimentary Canal

a. **Mouth** — the beginning of the simple digestive tract; external opening through which food is ingested

b. **Pharynx** — passageway from the mouth to the esophagus. Muscles here help move the food into the esophagus.

c. **Esophagus** — muscular passageway from the pharynx to the stomach. A flap of skin at the entrance to the esophagus called the *epiglottis* prevents food from entering the trachea.

d. **Stomach** — gastric juices continue the breakdown of food in this part of the simple digestive system.

e. **Small intestine** — food is further broken down for absorption here, as pancreatic juices and bile are added to the digesting food

f. **Caecum** — a blind gut; serves no function in swine but functions to help rabbits and horses digest forages

g. **Large intestine** — most moisture is squeezed from the undigested food here; also called the *colon*

h. **Rectum** — part of the alimentary canal where undigested food is held prior to release from the body

i. **Anus** — external opening of the digestive tract through which undigested food is expelled from the body

Accessory Organs

a. **Teeth** — serve to grind and pulverize food

b. **Salivary glands** — secrete digestive salivary juices (saliva), which begin the chemical breakdown of food

c. **Pancreas** — produces pancreatic digestive juices, which are released into the small intestine to help break down food

d. **Liver** — lobe-shaped organ that purifies the blood brought from the stomach, pancreas, and intestines. It also secretes bile, which is later added to the digestive juices in the small intestine.

e. **Gallbladder** — stores some bile from the liver and empties bile into the large intestine as waste to be removed from the body

❷ Wearing rubber gloves and safety goggles, spread out the digestive tract so that all parts can be easily seen. Identify the following parts and label them on the diagram in Observation 1.

a. From the mouth, feed passes through the esophagus and into the stomach. Locate the esophagus.

b. Follow the esophagus to the stomach. Trace the stomach toward the posterior of the tract until it joins the small intestine. The small intestine is the smaller tube that leads to the large intestine (the larger tube).

c. Where the small intestine and large intestine join is the caecum, a blind gut.

d. Follow the large intestine toward the posterior. There is a slight enlargement just before it reaches the anus. This is the rectum.

e. Find the liver. It is a lobed, dark red-brown organ.

f. Look for the gallbladder. It is a small, greenish sac embedded under the right lobe of the liver.

g. The pancreas is a small, pinkish, grainy organ found inside the bend made by the first section of the small intestine.

❸ Using a scalpel, remove a 2-inch section of the small intestine and cut it lengthwise. Wash its lining and examine the intestine section under a dissecting microscope. In Observation 2, draw what you see. The projections you see are called *villi*. They aid in the absorption of nutrients into the bloodstream from the small intestine.

144 ■ SECTION 8: ANIMAL SCIENCES

Observations

1. Label the parts of the simple digestive system on the following diagram.

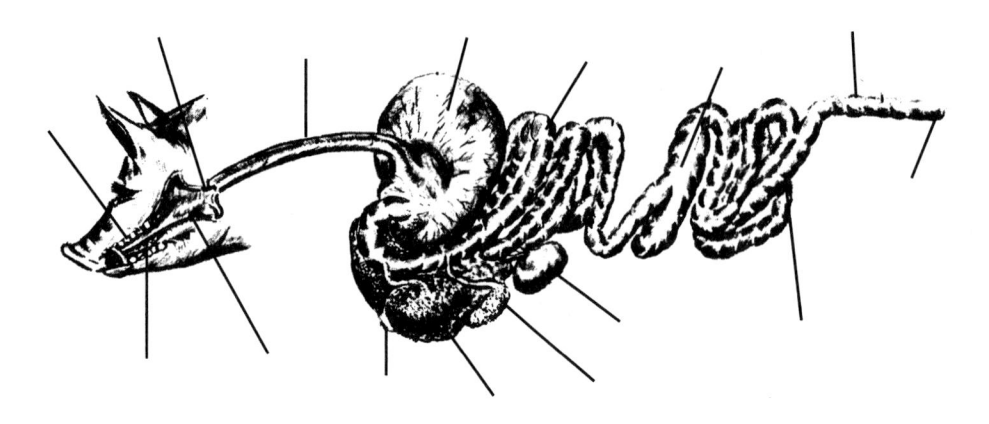

2. Draw the lining of a small intestine.

Conclusions

1. What is the function of the digestive system?

2. Identify the parts of the simple digestive system that function as follows:
   a. Squeezes moisture from the undigested food _____
   b. Produces bile _____
   c. Grinds and pulverizes the food _____
   d. Muscular passageway from the pharynx to the stomach _____
   e. Accessory organ that is found in the first bend of the small intestine and produces digestive juices to aid in digestion _____

3. Which part of the digestive system digests roughages in horses and rabbits?

4. What purpose do villi serve in the small intestine?

EXERCISE 38: DIGESTION IN RUMINANT ANIMALS

Materials Needed

- Digestive tract of a cow or sheep
- Rubber gloves
- Scalpel
- Four jars
- Dissecting microscope
- Compound microscope
- Rumen bacteria slide
- Personal Safety Set

SAFETY NOTE

Always wear eye protection and rubber gloves when handling dissection specimens. Wash hands before leaving laboratory.

Purpose To identify the parts and the functions of the parts of the ruminant digestive system.

Procedure

1. Study the following parts of the ruminant digestive system in Figure 26–8 of *Agriscience: Fundamentals & Applications, 3E* and locate them on the drawing.

Alimentary Canal

a. **Mouth** — the beginning of the simple digestive tract; external opening through which food is ingested

b. **Pharynx** — passageway from the mouth to the esophagus. Muscles here help move the food into the esophagus.

c. **Esophagus** — muscular passageway from the pharynx to the stomach. A flap of skin at the entrance to the esophagus called the *epiglottis* prevents food from entering the trachea.

d. **Rumen** — the first compartment of a ruminant's stomach. It is sometimes called the *paunch* and makes up approximately 80 percent of the ruminant stomach. It acts as a large storage vat where food is agitated and fermented, and where digestion begins.

e. **Reticulum** — this part of the ruminant stomach has a honeycomblike lining that can catch and hold foreign objects that may be picked up by the animal. For this reason, it is sometimes referred to as the *hardware stomach*. It composes approximately 5 percent of the ruminant stomach. It, along with the rumen, has many bacteria and protozoa that aid in the breakdown of forages (roughages).

f. **Omasum** — the third part of the ruminant stomach; grinds up food and may squeeze water from the food. It is sometimes called the *manyplies* and composes approximately 8 percent of the ruminant stomach.

g. **Abomasum** — this part of the ruminant stomach is sometimes called the *true stomach*. It works similarly to the stomach in the simple digestive tract in that digestive juices are produced here. The abomasum composes approximately 7 percent of the ruminant stomach.

h. **Small intestine** — food is further broken down for absorption here, as pancreatic juices and bile are added to the digesting food

i. **Caecum** — a blind gut; serves no function in ruminants but functions to help rabbits and horses digest forages

j. **Large intestine** — most moisture is squeezed from the undigested food here; also called the *colon*

k. **Rectum** — part of the alimentary canal where undigested food is held prior to release from the body

l. **Anus** — external opening of the digestive tract through which undigested food is expelled from the body

Accessory Organs

a. **Teeth** — serve to grind and pulverize food

b. **Salivary glands** — secrete digestive salivary juices (saliva), which begin the chemical breakdown of food

c. **Pancreas** — produces pancreatic digestive juices, which are released into the small intestine to help break down food

d. **Liver** — lobe-shaped organ that purifies the blood brought from the stomach, pancreas, and intestines. It also secretes bile, which is later added to the digestive juices in the small intestine

e. **Gallbladder** — stores some bile from the liver and empties bile into the large intestine as waste to be removed for the body

❷ Wearing rubber gloves and safety goggles, spread out the digestive tract so that all parts can be easily seen. Identify the following parts and label them on the diagram in Observation 1.

a. From the mouth, feed passes through the esophagus and into the rumen. Locate the esophagus.

b. Follow the esophagus to the rumen. Trace the rumen to the reticulum. There is no real division between the rumen and reticulum.

c. Next locate the omasum. Follow it until you find the abomasum.

d. Follow the abomasum until it joins the small intestine. The small intestine is the smaller tube that leads to the large intestine (the larger tube).

e. Where the small intestine and large intestine join is the caecum, a blind gut.

f. Follow the large intestine toward the posterior. There is a slight enlargement just before it reaches the anus. This is the rectum.

148 ■ SECTION 8: ANIMAL SCIENCES

❸ Label the jars "rumen," "reticulum," "omasum," and "abomasum."

❹ Cut the stomach open and take out some of the contents from each compartment. Put the contents in the jars labeled accordingly. Compare the digestion of food in each compartment. Record your findings in Observation 2.

❺ Cut out a section of the inner lining of each part of the stomach. Wash each section thoroughly and observe it under the dissecting microscope. Draw what you see in Observation 3.

❻ Place the rumen bacteria slide on the stage of the microscope and clip it down. Focus on low power and switch to high power. In Observation 4, draw the bacteria that you see.

Observations

1. Label the parts of the ruminant digestive system in the following diagram.

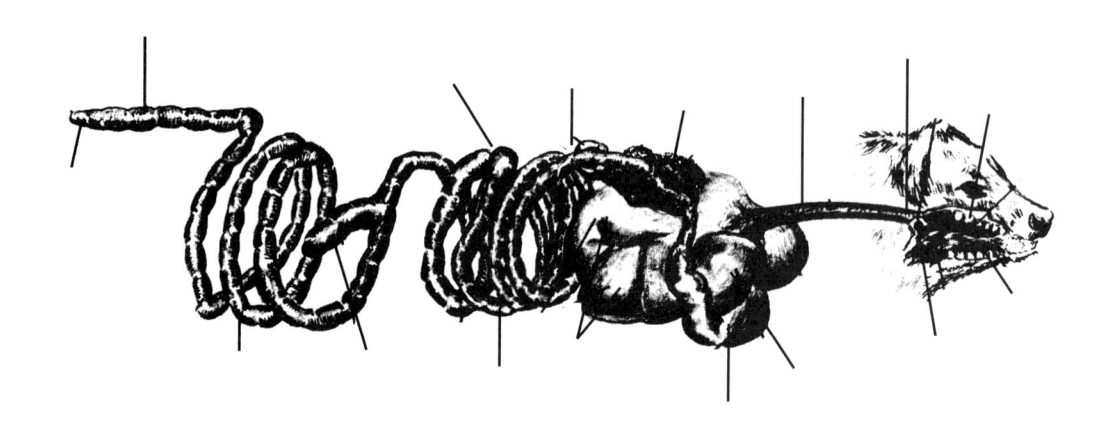

2. Describe the contents of each of the parts of a ruminant stomach.

   a. Rumen _____
   _____
   _____

   b. Reticulum _____
   _____
   _____

   c. Omasum _____
   _____
   _____

   d. Abomasum _____
   _____
   _____

3. Draw the inner lining of each part of the ruminant stomach as observed under the microscope.

   a. Rumen

   b. Reticulum

   c. Omasum

   d. Abomasum

4. Draw the bacteria as seen under the microscope.

Conclusions

1. Identify the function of each of the parts of the ruminant stomach.

   a. Rumen _____

   _____

   _____

   b. Reticulum _____

   _____

   _____

   c. Omasum _____

   _____

   _____

   d. Abomasum _____

   _____

   _____

2. What type of feed can ruminants digest that simple stomachs cannot?

   _____

   _____

3. How do bacteria help digestion in the ruminant stomach?

   _____

   _____

   _____

   _____

   _____

4. Which part of the ruminant stomach performs the same basic function as does the simple stomach?

   _____

   _____

5. Define the term *ruminant*.

   _____

   _____

   _____

   _____

EXERCISE 39: INTERNAL PARASITES

Purpose To study the structure and life cycles of several economically important internal parasites.

Materials Needed

- Sheep Liver Fluke Microscope Slide Set
- Sheep Tapeworm Microscope Slide Set
- Pig Roundworm Microscope Slide Set
- Compound microscope

Procedure

1. Study the information sheets for this exercise before continuing.
2. Examine the sheep liver fluke (fosciola hepatica life history) slides under the microscope. Draw the liver fluke eggs in Observation 1.
3. Examine the sheep liver fluke infected snail tissue under the microscope.
4. Examine the sheep liver fluke infected sheep liver under the microscope. Compare the two slides of infected tissue in Observation 2.
5. Examine the sheep tapeworm slides under the microscope. Draw the tapeworm eggs in Observation 3.
6. Examine the pig roundworm slides under the microscope. Draw the pig roundworm eggs in Observation 4.
7. Study the life cycle charts of the three internal parasites.

Observations

1. Draw the liver fluke eggs.

2. Compare the snail tissue damage to the liver damage. Draw or describe what you see.

3. Draw the tapeworm eggs.

151

152 ■ SECTION 8: ANIMAL SCIENCES

4. Draw the pig roundworm eggs,

5. Complete the following life cycle charts by drawing arrows to indicate the cycle flow.

   a.

   b.

c.

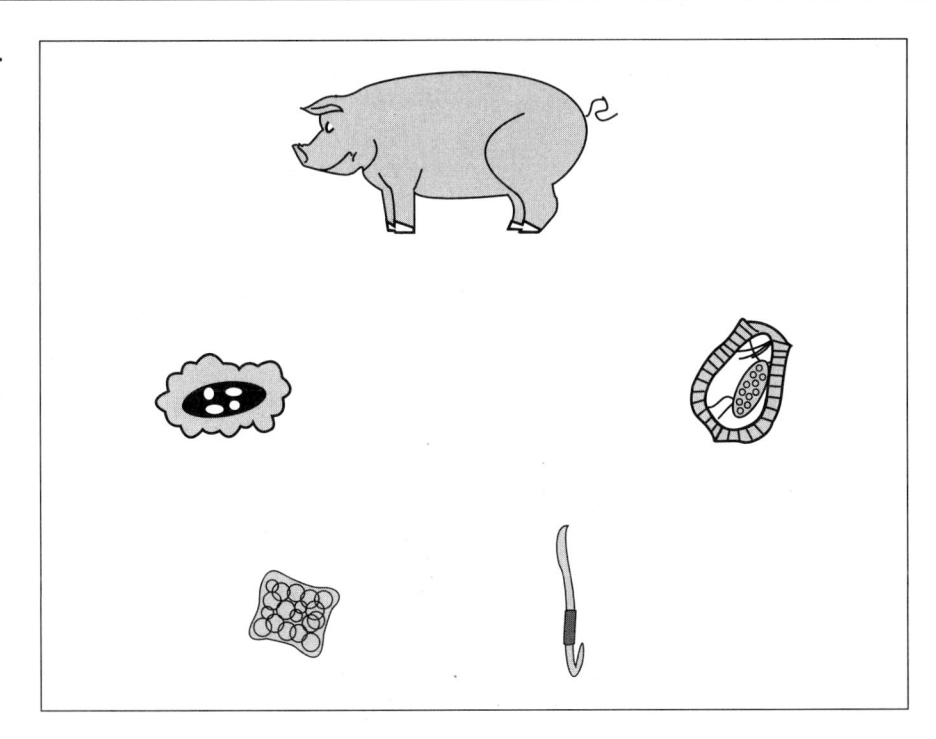

Conclusions

1. What type of damage does each of the following cause?

   a. Liver fluke _____

   b. Tapeworm _____

   c. Roundworm _____

2. How do these three internal parasites differ in the damage that they cause? How are they the same?

   _____
   _____
   _____
   _____

3. What differences are there in the life cycles of these parasites? What similarities are there?

   _____
   _____
   _____

4. How are most internal parasites controlled?

   _____
   _____
   _____

INFORMATION SHEET — LIVER FLUKE

The liver fluke is a parasite of cattle, sheep, goats, and humans. It is especially damaging to young animals.

The adult lives in the bile ducts, where eggs are laid and pass down into the intestines and out in the feces. The eggs must land in water to hatch. The larvae that hatch from these eggs swim about seeking a snail. The snail is necessary for completing the life cycle. The larvae develop for a period in the snail, then emerge and encyst on plants along the water. Livestock eat the water plants and become infected. The young flukes pass to the intestines and burrow through the abdominal cavity and into the liver, where they live principally on blood. Egg production begins approximately three months after entering the animal.

The fluke causes irritation, thickening of the bile duct, and fibrosis of the liver, making the animal unfit for human consumption.

The symptoms of liver flukes are anemia and weight loss. Highly infested animals may die.

Rotation of pastures and the use of water troughs help in control. Chemical treatment will kill the adult flukes in the animal. Control of snails will break the cycle but is hard to do.

INFORMATION SHEET — TAPEWORM

There are several species of tapeworm that affect livestock. The broad tapeworm affects all classes of livestock and humans.

The adult lives in the small intestines, where it may reach a length of 10 or more feet. Segments of the tapeworm that contain eggs are continuously breaking off and passing out in the feces. The eggs are eaten by the orbatid mite, which lives in the grass and weeds and serves as an intermediate host. The eggs develop in these mites, then are eaten by livestock and hatch in the small intestines. They feed on animal foodstuff and grow to maturity. There is no physical damage to the host; rather tapeworms compete with the host for food.

Symptoms of tapeworms include unthriftiness, loss of weight, diarrhea, and emaciation.

Tapeworms can be eliminated from their hosts via the use of chemicals.

INFORMATION SHEET — ROUNDWORM

Roundworms are the most important group of internal parasites from an economic standpoint. There are many types of roundworm, which affect almost every type of livestock. The roundworms of greatest concern are found in the digestive system, mostly in the stomach and intestines.

Roundworms in the adult stage live as blood-sucking parasites attached to the stomach wall. The eggs pass from the host to the feces and hatch into larvae in 15 to 20 days, depending on the temperature and humidity. The larvae crawl up blades of grass, are eaten by the animal, and travel to the stomach lining until they mature.

In the process of penetrating the stomach lining before maturing, they cause severe damage by reducing the digestion of nutrients by the host and producing poisons. Young, undernourished, or diseased animals are hardest hit.

The most common symptom is anemia, which can be indicated by a paleness of the gums and the whites of the eyes. In light infestations, the animal will have a dull hair coat, an unthrifty appearance, and will sometimes scour. In severe infestations there will be persistent scouring, weight loss, anemia, prostration, and sometimes death.

Sanitation and pasture rotation are good control measures. Chemical dewormers are used in treating infested animals.

EXERCISE 40:

CONTROLLING DISEASES VIA ANTIBIOTICS

Materials Needed
- ✔ Antibiotics Effects Kit
- ✔ Marking pen

Purpose   To determine the effects of antibiotics on different organisms.

Procedure

1 Study the general aseptic techniques found in the antibiotics effects kit.

2 Read the Introduction, Objectives, and About the Experiment sections found in the instruction sheet.

3 Follow the aseptic technique to pour the BHI plates.

4 Rehydrate and dilute the cultures of bacteria and fungus.

5 Prepare the control plate.

     a. Using a marking pen, divide the bottom of one BHI plate into three equal sections. Label each section with a code for the name of the microorganism.

     b. Follow the aseptic technique instructions and inoculate the plate by squeezing one drop of the appropriate microorganism onto the designated portion of the agar. The drop should be at the center edge of the division.

     c. Immediately close the cover. Do not swirl the petri dish.

6 Prepare a fleming plate.

     a. Using two BHI plates, label each with a code for one of the bacterial species.

     b. Follow the aseptic technique instructions and inoculate each plate with the designated organism, using one drop of the organism and spreading it evenly over the plate. Allow to dry.

     c. Use the pipette for the penicillium mold. Make sure the neck of the pipette is full of material. Carefully push the tip of the pipette into the agar and very gently squeeze one drop of material into the center of the agar plate. Close the plate.

     d. Using scissors, cut off the tip of the pipette and repeat the procedure on the other plate.

SECTION 8: ANIMAL SCIENCES ■ 157

7 Prepare the antibiotic plates

a. Label four BHI plates with the code for one bacterial species and four with the code for the other bacterial species.

b. Follow the aseptic technique and inoculate each plate, using one drop of the designated organism and spreading it uniformly over the plate. Allow to dry. Briefly flame the forceps.

c. Carefully remove a disk from one of the three disk containers.

d. Gently place the disk on top of the agar in the center of the plate. Cover the plate. Flame the forceps.

e. Repeat steps 7c and 7d for each of the BHI plates so that you have one plate for each antibiotic type on each type of organism and one control organism of each type. Cover all plates.

f. Label the plates with the proper antibiotic codes and the organism codes.

8 Incubate the plates upside down to keep contamination (from condensation, improper handling, etc.) to a minimum.

9 Observe the dishes at 5 and 10 days. Record your observations in the table in Observation 1.

Observations

1. Record observations in the following table.

	Tetracycline Code _____	Erythromycin Code _____	Penicillin Code _____	Control Code _____
Micrococcus Roseus Code _____				
Escherichia Coli Code _____				

Conclusions

1. How does the growth in the sensitivity plates compare to that in the control plates?

2. Describe the condition of the fleming plate.

3. How do antibiotics aid farmers and ranchers in the livestock industry?

4. Which bacterium was the most resistant to the antibiotic agents?

EXERCISE 41: UNDERSTANDING GENETICS

Materials Needed
- ✔ Pencil or pen
- ✔ Calculator

Purpose To provide students with an understanding of how different gene combinations result in different individuals.

Procedures

1. Study the following terms related to genetics.

   a. **Heredity** — the passing on of traits or characteristics from parents to offspring

   b. **Genes** — components of cells that determine the individual characteristics of living things

   c. **Genetics** — the study of heredity

   d. **Genotype** — what the genes look like

   e. **Phenotype** — physical appearance of an individual

   f. **Dominant** — gene that expresses itself to the exclusion of other genes

   g. **Recessive** — gene that expresses itself only in the absence of a dominant gene

   h. **Heterozygous** — pairs of genes that are different

   i. **Homozygous** — pairs of genes that are alike

   j. **Incomplete dominance** — neither gene expresses itself to the exclusion of the other

2. Complete the squares for the mating of the following two animals: Bb—Black bull and bb—Red cow   Bb × bb (Observation 1)

3. Complete the squares for the mating of the following two animals: PP—Polled bull and pp—Horned cow   PP × pp (Observation 2)

4. Complete the squares for the mating of the following two animals: BbPP—Black polled bull and bbpp—Red horned cow   BbPP × bbpp (Observation 3)

5. Complete the squares for the mating of the following two animals: RR—Red Shorthorn bull and WW—White Shorthorn cow   RR × WW (Observation 4)

159

Observations

1. Complete the squares that follow:

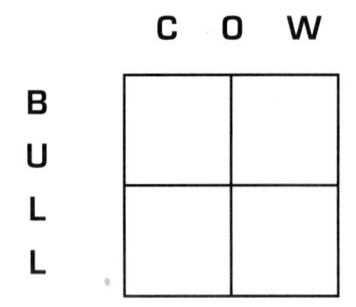

```
       C   O   W
   B ┌───┬───┐
   U │   │   │
   L ├───┼───┤
   L │   │   │
     └───┴───┘
```

2. Complete the squares that follow:

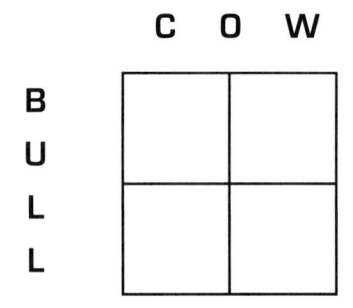

```
       C   O   W
   B ┌───┬───┐
   U │   │   │
   L ├───┼───┤
   L │   │   │
     └───┴───┘
```

3. Complete the squares that follow:

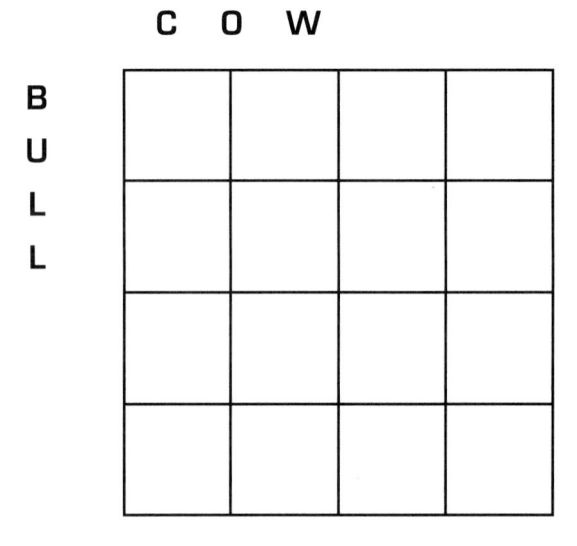

```
        C    O    W
   B ┌────┬────┬────┬────┐
   U │    │    │    │    │
   L ├────┼────┼────┼────┤
   L │    │    │    │    │
     ├────┼────┼────┼────┤
     │    │    │    │    │
     ├────┼────┼────┼────┤
     │    │    │    │    │
     └────┴────┴────┴────┘
```

4. Complete the squares below:

```
       C   O   W
   B ┌───┬───┐
   U │   │   │
   L ├───┼───┤
   L │   │   │
     └───┴───┘
```

Conclusions

1. Define the following terms:

    a. Dominant _____

    _____

    b. Gene _____

    _____

    c. Genetics _____

    _____

    d. Genotype _____

    _____

    e. Heredity _____

    _____

    f. Heterozygous _____

    _____

    g. Homozygous _____

    _____

    h. Incomplete dominance _____

    _____

    i. Phenotype _____

    _____

    j. Recessive _____

    _____

2. In Observation 1, identify the following:

    a. The dominant characteristic? _____

    _____

    b. The recessive characteristic? _____

    _____

    c. The phenotypic ratio _____

    _____

    d. The genotypic ratio _____

    _____

3. In Observation 2, identify the following:

    a. The dominant characteristic? _____

    _____

162 ■ SECTION 8: ANIMAL SCIENCES

    b. The recessive characteristic? _____

    c. The phenotypic ratio _____

    d. The genotypic ratio _____

4. In Observation 3, identify the following:

    a. The dominant characteristic? _____

    b. The recessive characteristic? _____

    c. The phenotypic ratio _____

    d. The genotypic ratio _____

5. In Observation 4, identify the following:

    a. The dominant characteristic? _____

    b. The recessive characteristic? _____

    c. The phenotypic ratio _____

    d. The genotypic ratio _____

EXERCISE 42:

INJECTION PROCEDURES

Purpose To practice procedures for properly vaccinating animals for disease prevention and control.

- ✔ Two or three 10-ml syringes
- ✔ ½-inch needles
- ✔ 1-inch needles
- ✔ Three containers of water
- ✔ Food dye—red, green, blue
- ✔ Three or four grapefruit
- ✔ Ruler

Procedures

1 Study the following terms related to injection procedures.

  a. **Injection** — the process of administering drugs by needle and syringe

  b. **Antibiotic** — substance used to help prevent or control certain diseases of animals

  c. **Intradermal** — between layer of skin

  d. **Intramuscular** — in a muscle

  e. **Intraperitoneal** — in the abdominal cavity

  f. **Intrarumenal** — in the rumen

  g. **Intravenous** — in a vein

  h. **Subcutaneous** — under the skin

  i. **Syringe** — an instrument used to give injections of medicine or to draw body fluids from animals

  j. **Vaccination** — the injection of an agent into an animal to prevent disease

2 Add blue dye to one water container, red dye to a second water container, and green dye to the third. Add dye to the water until the water is a dark color.

3 Using the ½-inch needle, inject one grapefruit with 2 milliliters of green-colored dye directly under the skin of the grapefruit (subcutaneous).

4 Using a 1-inch needle, inject the same grapefruit used in #3 with 2 milliliters of red-colored dye directly into the grapefruit (intramuscular).

5 Using the ½-inch needle, inject one grapefruit with 5 milliliters of blue-colored dye directly under the skin of the grapefruit (subcutaneous).

6 Using a 1-inch needle, inject the same grapefruit used in Step 5 with 5 milliliters of green-colored dye directly into the grapefruit (intramuscular).

7 Using the ½-inch needle, inject one grapefruit with 10 milliliters of red-colored dye directly under the skin of the grapefruit (subcutaneous).

8 Using a 1-inch needle, inject the same grapefruit used in Step 7 with 10 milliliters of blue-colored dye directly into the grapefruit (intramuscular).

9 Wait 10 minutes then cut each grapefruit along a line where the intramuscular injection was given. Record your findings in Observation 1.

10 Peel the grapefruit around the area where the subcutaneous injection was given. Record your findings in Observation 2.

Observations

1. Provide the information requested in the following for the intramuscular injection.

Intramuscular Injection	Grapefruit 1 2 ml	Grapefruit 2 5 ml	Grapefruit 3 10 ml
Approximate Size of Area Colored with Dye			

2. Provide the information requested in the following for the subcutaneous injection.

Subcutaneous Injection	Grapefruit 1 2 ml	Grapefruit 2 5 ml	Grapefruit 3 10 ml
Approximate Size of Area Colored with Dye			

Conclusions

1. Which type of injection distributed the dye to the largest area?

2. Was the amount of area covered with the dye proportional from the smallest to the largest? Give reasons why it was or was not.

3. Which type of injection was easier for you to perform? Why?

4. Besides intramuscular and subcutaneous, list other types of injections that may be given.

SECTION: 9

Food Science and Technology

EXERCISE 43: FOOD NUTRIENTS

Purpose   To observe the different food nutrients found in the foods we eat.

Materials Needed

- Twenty test tubes
- Test tube rack
- Five eyedroppers
- Hot plate
- Sudan IV
- Buiret reagent
- Potassium iodide—Iodine solution
- Benedict's solution
- Milk
- Cream or butter
- Syrup
- Canned whole corn
- Distilled water

Procedure

1 Group the test tubes in the test tube rack in groups of four. Label the five different groups as follows:

   a. Glucose — sugar
   b. Starch
   c. Fat
   d. Protein
   e. Water

2 Put 5 milliliters of water into each of the test tubes.

3 Add the following to the test tubes:

   a. Three to five drops of syrup to the four test tubes marked "glucose."
   b. Three to five drops of juice from the canned corn to the test tubes marked "starch."
   c. Three to five drops of cream or melted butter to the four test tubes marked "fat."
   d. Three to five drops of milk to the four test tubes marked "protein."
   e. Three to five drops of distilled water to the four test tubes marked "water."

4 Add 10 drops of Benedict's solution to the first tube of each group of four test tubes. Heat the four tubes gently in a hot water bath for 10 minutes. If sugar is present, the color will change to orange. Record your findings in the table in Observation 1.

5 Add one drop of potassium iodide-iodine solution to the second tube in each group of four test tubes. Starch turns blue in the presence of iodine. Glycogen (animal starch) turns mahogany red in the presence of iodine. Record your findings in the table in Observation 2.

SECTION 9: FOOD SCIENCE AND TECHNOLOGY ■ 169

6 Add one drop of Sudan IV to the third tube in each group of four test tubes. Sudan IV is an oil-soluble dye and will tag oil globules. Record your findings in the table in Observation 3.

7 Add 10 drops of Buiret reagent to the last test tube in each group of four test tubes. Buiret reagent changes from blue to violet pink if peptides (protein building blocks) are present. The color may fade. Allow 10 to 15 minutes for the reaction to take place. Record your findings in the table in Observation 4.

Observations

1. Sugar test results

Test Tube	Reaction (if any)
Syrup	
Corn Juice	
Cream	
Milk	
Water	

2. Starch test results

Test Tube	Reaction (if any)
Syrup	
Corn Juice	
Cream	
Milk	
Water	

170 ■ SECTION 9: FOOD SCIENCE AND TECHNOLOGY

3. Fat test results

Test Tube	Reaction (if any)
Syrup	
Corn Juice	
Cream	
Milk	
Water	

4. Protein test results

Test Tube	Reaction (if any)
Syrup	
Corn Juice	
Cream	
Milk	
Water	

Conclusions

1. Which foods reacted to the starch test?

2. Which foods showed sugar present?

3. What were the protein foods?

4. Which foods reacted to the fat test?

5. What reactions took place in the distilled water?

6. List the foods (if any) that showed more than one of the food nutrients and which nutrients were present.

EXERCISE 44:

FOOD PRESERVATIVES

Materials Needed

- ✔ Bread
- ✔ Swiss cheese
- ✔ Hot plate
- ✔ Two 20-ml test tubes
- ✔ Two 250-ml beakers
- ✔ Distilled water
- ✔ Test tube holder
- ✔ Sulfuric acid solution
- ✔ Funnel
- ✔ Filter paper
- ✔ Spatula
- ✔ Ethyl alcohol
- ✔ Pencil

Purpose To determine the presence of food additives in the foods that we eat.

Procedure

1 Place 400 milliliters of distilled water into a 600-ml beaker and bring to a low boil on the hot plate.

2 Cut the Swiss cheese into small squares and place in a 250-ml beaker. Add 100 milliliters of distilled water.

3 Mash the cheese and stir with a spatula to mix the cheese and water as much as possible. Allow to sit while preparing the bread.

4 Place a slice of bread in a 250-ml beaker and add 100 milliliters of distilled water. Let the bread soak. Stir the mixture with a spatula to mix the bread and water as much as possible.

5 Using the funnel with filter paper, filter approximately 10 milliliters of the cheese-water mixture in one test tube.

6 Follow the same procedure for the bread mixture.

7 Add 5 milliliters of sulfuric acid solution to both the Swiss cheese and bread solutions. Record the reaction in Observation 1.

8 Add 3 milliliters of ethyl alcohol to each test tube and place in water bath on hot plate.

9 When the solutions begin to boil, carefully observe the odor coming from the vapors. In Observation 2, record the odor you detect.

NOTE: An odor similar to pineapple indicates Ethyl propionates.

Observations

1. What happened to the cheese solution when the sulfuric acid was added?

2. What odor did you detect when you heated the cheese solution? The bread solution?

Conclusions

1. Why are preservatives necessary in foods such as cheese and bread?

2. What are some natural preservatives found in food?

3. If food preservatives were not used, how do you think it would affect food supply and prices?

EXERCISE 45: PRODUCING DAIRY PRODUCTS

Materials Needed

- ✔ 12-inch piece of cheesecloth
- ✔ Hot plate
- ✔ Plastic cups— 16 ounces each
- ✔ Cooking thermometer
- ✔ Milk
- ✔ Salt
- ✔ Rennin or rennilase

Purpose To demonstrate the process involved in producing cheese.

Procedure

1. Heat the milk to 88–90°F on the hot plate. Use the thermometer to monitor the temperature.

2. Fill the plastic cups approximately two-thirds to three-fourths full.

3. Immediately add 12 drops of the rennin solution to the milk. Stir continuously until the rennin enzyme is completely mixed in.

4. Let the milk set undisturbed until the curd (solid portion) has separated from the whey (liquid portion).

5. Filter the liquid from the solid by folding the cheesecloth in half and placing over the top of the cup to act as a strainer.

6. Pour the liquid into another plastic cup.

7. Salt the cheese (curds) and allow to set further to harden.

8. Cut and eat.

Observations

1. Describe the color of the curds. Describe the color of the whey.

2. Describe the taste of the finished cheese.

Conclusions

1. What function does the enzyme rennin serve in this experiment?

2. How do you explain the different tastes of different cheeses?

SECTION: 10

Communications and Management in Agriscience

EXERCISE 46: UNDERSTANDING DIMINISHING RETURNS

Materials Needed
- Pencil
- Calculator

Purpose To provide an understanding of the law of diminishing returns.

Procedure

1 Study the terms associated with diminishing returns:

a. **Marginal** — refers to incremental changes, increases or decreases, which occur at the edge or margin (additional).

b. Δ — shorthand for "change in."

c. **Total physical product (TPP)** — the total output or yield.

d. **Average physical product** — average amount of output produced by each unit of input at each input level.

$$APP = \frac{\text{Total physical product (TPP)}}{\text{Input level}}$$

e. **Marginal physical product (MPP)** — the additional or extra total physical product produced by using an extra unit of input.

$$MPP = \frac{\Delta \text{ total physical product}}{\Delta \text{ input level}}$$

f. **Marginal value of product (MVP)** — the additional or marginal income received from using an additional unit of input.

$$MVP = \frac{\Delta \text{ total value of product (TVP)}}{\Delta \text{ input level}}$$

g. **Total value of product (TVP)** — the product of multiplying quantity of output (TPP) by selling price.

h. **Marginal input cost (MIC)** — the change in total input cost or the addition to total input cost caused by using an additional unit of input.

$$MIC = \frac{\Delta \text{ total input cost}}{\Delta \text{ input level}}$$

i. The profit maximizing point is the input level where MVP = MIC.

SECTION 10: COMMUNICATIONS AND MANAGEMENT IN AGRISCIENCE ■ 179

2 Study the following examples:

Input price = $10.00 per unit
Output price = $2.50 per unit

Input Level	TPP	APP	MPP	TVP ($)	MVP ($)	MIC ($)
0	0	—	—	0	—	—
1	10	10	10	25	25	10
2	28	14	18	70	45	10
3	42	14	14	105	35	10
4	52	13	10	130	30	10
5	60	12	8	150	20	10
6	66	11	6	165	15	10
7	70	10	4	175	10	10
8	72	9	2	180	5	10
9	70	7.78	−2	175	−5	10
10	68	6.80	−4	170	−10	10

a. Input level and total physical product (TPP) are given. Input price and output price are given.

b. Study the **a**verage **p**hysical **p**roduct (APP) column.

Examples:

APP = TPP/input = 10/1 = 10 for level 1

APP = TPP/input = 28/2 = 14 for level 2

c. Study the **m**arginal **p**hysical **p**roduct (MPP) column.

Examples:

MPP = TPP/Δ input = 10 − 0/1 − 0 = 10 for level 1

MPP = TPP/Δ input = 28 − 10/2 − 1 = 18 for level 2

d. Study the **t**otal **v**alue of **p**roduct (TVP) column.

Examples:

TVP = TPP × selling price = 10 × $2.50 = $25.00 for level 1

TVP = TPP × selling price = 28 × $2.50 = $70.00 for level 2

e. Study the **m**arginal **v**alue of **p**roduct (MVP) column.

Examples:

MVP = TVP/Δ input = 25 – 0/1 – 0 = $25.00 for level 1

MVP = TVP/Δ input = 70 – 25/2 – 1 = $45.00 for level 2

f. Study the marginal input cost (MIC) column.

Examples:

MIC = total input cost/Δ input = 1 – 0 ($10.00)/1 –0 = 10/1 = 10

MIC = total input cost/Δ input = 2 – 1 ($10.00)/2 – 1 = 10/1 = 10

g. The profit maximizing point is where MVP = MIC. If you look at input level 7, you find MVP = 10 and MIC = 10.

h. The Law of Diminishing Returns states that:

As additional units of a variable input are used in combination with one or more fixed inputs, the marginal physical product will eventually begin to decline. Study the MVP column on the chart.

SECTION 10: COMMUNICATIONS AND MANAGEMENT IN AGRISCIENCE

Observations

Complete the following chart.

Fertilizer costs $15.00 per unit.
Corn sells for $3.00 per bushel.

Units of Fertilizer	TPP (Corn)	APP	MPP	TVP ($)	MVP ($)	MIC ($)
0	70	—	—	0	—	—
1	96					
2	118					
3	129					
4	136					
5	140					
6	150					
7	140					
8	136					

1. Complete the **average physical product (APP)** column.

2. Complete the **marginal physical product (MPP)** column.

3. Complete the **total value of product (TVP)** column.

4. Complete the **marginal value of product (MVP)** column.

5. Complete the **marginal input cost (MIC)** column.

182 ■ SECTION 10: COMMUNICATIONS AND MANAGEMENT IN AGRISCIENCE

Conclusions

1. What is the profit-maximizing amount of fertilizer to use on corn?

2. If the TVP column is higher on level 6, why is it not the profit-maximizing level?

3. If fertilizer cost only $12.00 per unit, what would be the profit-maximizing level of fertilizer to use? Why?

EXERCISE 47: UNDERSTANDING INTEREST AND CREDIT

Purpose To demonstrate the different methods of charging interest for borrowing money.

Materials Needed
- ✔ Pencil
- ✔ Calculator

Procedure

1 Study the following terms:

a. **Principal** — the amount of money received from a loan transaction.

b. **Interest** — the charge that is paid for use of money borrowed from an institution or person.

c. **Interest rate** — the charge paid for money borrowed, expressed as a percentage of the principal.

d. **Time** — the length of payback for money borrowed, usually expressed in terms of years.

e. **True or actual annual rate** — a standard or basic interest rate for comparison of different methods of charging interest.

f. **Simple interest** — the interest rate on a loan with a single (one) payment.

g. **Interest on the unpaid balance** — a method of charging interest that causes the interest payments to decline as the principal loan balance declines.

h. **Discounted interest** — a type of loan that requires the interest to be paid in advance or at the time the loan is received. The interest is deducted from the original amount of the loan, leaving the borrower with less than the amount actually borrowed.

i. **Add-on-interest** — usually charged on installment loans with interest charged on the full amount of the loan for the entire time of the loan.

2 Study the following examples:

Simple Interest

a. $12,000 is borrowed for one year at 10 percent interest.

Formula: $I = PRT$

$I = \$12,000 \times 10\% \times 1 \text{ (yr)} = \$1,200$

The simple interest rate is also the true annual rate: 10%

b. $12,000 is borrowed at 10% interest, $6,000 will be repaid at the end of six months, and the remaining $6,000 will be paid at the end of the year.

I = PRT

I = $12,000 × 10% × ½ (year) = $600

I = $ 6,000 × 10% × ½ (year) = $300

$900

When interest is paid only on the unpaid balance, the stated interest rate is the same as the true annual rate: 10%.

Discounted Interest

c. $12,000 is borrowed at 10 percent but is discounted so $1,200 interest is deducted.

$$\text{rate} = \frac{\text{Interest}}{\text{Adjusted principal} \times \text{time}} \times 100$$

$$= \frac{\$1,200}{(\$12,000 - 1,200) \times 1}$$

$$= \frac{\$1,200}{\$10,800 \times 1}$$

$$= 11.11\%$$

d. $12,000 is borrowed at 10% for 3 years with 36 equal payments.

I = PRT = $12,000 × 10% × 3 (years) = $3,600

$$\text{true rate} = \frac{2 \times \text{no. of payments} \times \text{interest charged}}{\text{Principal} \times \text{time} \times (\text{no. of payments} + 1)}$$

$$= \frac{2 \times 36 \times 3,600}{12,000 \times 3 \times 37}$$

$$= \frac{259,200}{1,332,000}$$

$$= 19.46\%$$

e. Chart 1

The chart on the following page shows the difference in interest paid and true annual interest rate for different types of interest charged on the following.

Amount Borrowed = $12,000
Interest Rate = 10%

SECTION 10: COMMUNICATIONS AND MANAGEMENT IN AGRISCIENCE ■ 185

Kind of Interest	Rate	Time (Years)	Interest Paid	True Annual Interest Rate
Simple Interest	10%	1	$1,200	10%
Interest on Unpaid Balance	10%	1	$ 900	10%
Discounted Interest	10%	1	$1,200	11.11%
Add-on Interest	10%	3	$3,600	19.46%

Observations

1. You are planning to borrow $15,000. You have been offered a 12 percent rate with the following options:

   a. Simple interest for one year.

   b. Interest on unpaid balance for one year with half due in 6 months and half due at the end of the year.

   c. A discounted loan for one year.

   d. Add-on-interest for three years with 36 payments.

   Calculate the amount of interest and true APR for each of the preceding options and record them in the following chart.

Kind of Interest	Rate	Time (Years)	Interest Paid	True Annual Interest Rate
Simple Interest	12%	1		
Interest on Unpaid Balance	12%	1		
Discounted Interest	12%	1		
Add-on Interest	12%	3		

186 ■ SECTION 10: COMMUNICATIONS AND MANAGEMENT IN AGRISCIENCE

Conclusions

1. What is the lowest interest rate? What is the highest?

2. If you were to make payments each month or put in the bank an amount equal to a payment (for those due in one lump sum), which would cost the most per month? Which would cost the least?

3. Based on the answer to the previous conclusion 2, why do you think some people would choose an add-on-interest loan?

EXERCISE 48: PREPARING BALANCE SHEETS

Materials Needed
- ✔ Pencil
- ✔ Calculator

Purpose To demonstrate the use of balance sheets in agribusiness.

Procedure

1 Study the following terms:

a. **Balance sheet** — a form showing the financial condition of a business at a definite point in time. It lists all assets, value of assets, and liabilities of the business. It is also known as a *net worth statement, financial statement*, or *statement of financial condition*.

b. **Assets** — the property or resources owned and controlled by a business.

c. **Current assets** — cash or other assets that can be converted to cash within 12 months. Examples: checking and/or savings accounts, accounts receivable, inventory held for sale (market animals and stored crops), other near-cash items such as securities, stocks and bonds, and cash value of life insurance.

d. **Intermediate assets** — resources or production items with useful lifetimes of 1 to 10 years. Examples: equipment, machinery, and breeding stock. These assets are generally depreciable.

e. **Fixed assets** — also called *long-term assets*; include permanent items such as real estate and improvements on buildings. The useful lifetime is generally over 10 years.

f. **Liabilities** — all of the debt obligations of the business

g. **Current liabilities** — debts due within the operating year, that is, a 12-month period. Examples: notes and accounts payable, rents, taxes, interest, and principal payments due on intermediate or long-term debt within the next 12 months.

h. **Intermediate liabilities** — non-real-estate debt that corresponds to intermediate assets. Loan terms are normally for a period of 1 to 10 years. Examples: improvements to real estate, equipment purchases, breeding livestock and dairy stock, and capital requirements for major adjustments in farm operation.

i. **Long-term liabilities** — mortgages and land contracts on real estate minus principal due within 12 months.

j. **Net worth**—total assets minus total liabilities; also called *owner's equity*.

k. **Liquidity**—the ability of a business to make enough cash to pay bills without disrupting business. One of the best measures of liquidity is current ratio.

l. **Current ratio** $= \dfrac{\text{Current assets}}{\text{Current liabilities}}$

m. **Intermediate ratio**—the intermediate liquidity of the business can be determined by the intermediate ratio.

$$\text{intermediate ratio} = \dfrac{\text{Current + intermediate assets}}{\text{Current + intermediate liabilities}}$$

n. **Solvency**—the ability to pay debts; shows the amount that would be left after all assets are converted to cash and all debts paid. Determined by:

$$\text{net capital ratio} = \dfrac{\text{Total assets}}{\text{Total liabilities}}$$

o. **Equity**—the owner's share of the business. Shown by debt-equity ratio; also known as the *debt-to-net worth ratio*.

$$\text{(debt-to-net worth ratio)} = \dfrac{\text{Total liabilities}}{\text{Net worth (owner's equity)}}$$

Lenders are interested in this ratio because they usually prefer to provide loans that are equal to less than 50 percent of assets. Lenders prefer a debt-equity ratio of less than 1.

Current Assets = $25,000   Current Liabilities = $15,000
Intermediate Assets = $75,000   Intermediate Liabilities = $50,000
Fixed Assets = $200,000   Long-Term Liabilities = $150,000

---

❷ Using the information in the preceding chart:

a. Figure the current ratio.

b. Determine the intermediate ratio

c. Use the net capital formula to determine the net capital ratio.

d. Determine the net worth.

e. Determine the debt-equity ratio.

SECTION 10: COMMUNICATIONS AND MANAGEMENT IN AGRISCIENCE

Observations

1. Use the following information to complete the balance sheet.

Value of machinery	$ 75,000.00
Loan on feeder steers	$ 18,000.00
Value of cow herd	$ 21,000.00
Mortgage on land	$125,000.00
Checking account balance	$ 2,700.00
Value of land	$337,000.00
Bank note due next month	$ 16,500.00
Balance on machinery loan	$ 34,000.00
Value of feeder steers	$ 29,000.00
Value of stored grain	$ 23,00.00

   a. Fill in the current assets.

   b. Fill in the current liabilities

   c. Complete the intermediate assets.

   d. Complete the intermediate liabilities.

   e. What are the fixed assets? List them on the balance sheet.

   f. Add the long-term liabilities to the sheet.

   g. Determine the net worth.

BALANCE SHEET

Current Assets

_____

_____

_____

Current Liabilities

_____

_____

_____

Intermediate Assets

_____

_____

Intermediate Liabilities

_____

_____

Fixed Assets

_____

Long-Term Liabilities

_____

Total Assets _____

Total Liabilities _____

Net Worth _____

Conclusions

1. What is the current ratio?

2. What is the intermediate ratio?

3. What is the net capital ratio?

4. What is the debt-equity ratio?

5. Is the business solvent?

EXERCISE 49: PROFIT-LOSS STATEMENTS

Materials Needed
- ✔ Pencil
- ✔ Calculator

Purpose To demonstrate the use of profit-loss statements in agribusiness.

Procedure

1 Study the following terms:

a. **Profit and Loss (income) statement** — the financial record that reflects the profitability of the business over a specified period of time

b. **Income** — cash income includes money received from the sale of crops, livestock, and livestock products during the year. It also includes custom work and government payments. Gain from the sale of capital assets is also entered as cash income; the entry is the difference between the selling price and the book value. Noncash income includes products used in the home or given to a farm for custom work. Positive changes in inventory are also considered noncash income.

c. **Expenses** — cash expenses are monies paid to operate the business, such as those for seed, feed, fertilizer, market livestock, and fuel. Fixed-cash expenses include taxes, interest, insurance, and, in some cases, repairs. Noncash expenses include depreciation on machinery, equipment, buildings, and purchased breeding stock. Negative changes in inventory are recorded as noncash expenses. Not recorded as expenses are costs of any new capital assets and principal payments on loans.

d. **Cash accounting method** — income and expenses are recorded only when actual cash transactions occur. Most noncash income and expenses are not recorded under the cash method. Depreciation is an exception, as is the value of farm products used in the home. Both of these are included under the cash method.

e. **Accrual method** — records income when it is earned and expenses when they are incurred. Inventory changes are included in the accrual method. Because of this, it is actually a more accurate method of accounting.

f. **Net cash income** — the difference between total cash income and total cash expenses

g. **Net farm income** — net cash income plus noncash adjustments

h. **Return to capital** or **return on investment** — a measure of profitability based on a ratio obtained by dividing the return to total capital by total farm assets

i. **Return to capital** — adjusted net farm income minus opportunity cost of labor minus opportunity cost of management

j. **Rate of return to capital (%)** $= \dfrac{\text{Return to total capital}}{\text{Total farm assets}} \times 100$

k. **Adjusted net farm income** — net farm income plus interest paid minus value of unpaid family labor

l. **Return to labor and management** — that portion of net farm income that remains to pay the owner for personal labor and management after capital is paid a return equal to its opportunity cost. Return to labor and management equals adjusted net farm income minus opportunity cost on total capital.

m. **Return to labor** — return to labor and management minus opportunity cost of management

n. **Return to management** — return to labor and management minus opportunity cost of labor

o. **Return to equity** — net farm income minus opportunity cost of own labor minus opportunity cost of management minus value of unpaid family labor

p. **Rate of return to equity (%)** $= \dfrac{\text{Return to equity}}{\text{Net worth}} \times 100$

2 Working with formulas:

Net Farm Income	$ 80,000.00
Value of Operator's Management	$ 5,000.00
Owner's Equity	$500,000.00
Interest Paid	$ 25,000.00
Value of Total Assets	$750,000.00
Value of Operator's Labor	$ 15,000.00
Opportunity Cost of Capital	10%

a. What is the return to capital in percent?

b. What is the return to management?

c. What is the return to equity in percent?

Observations

1. Use the following information to complete the profit and loss statement:

Grain sorghum sales	$30,000.00
Alfalfa hay sales	$ 9,500.00
Building depreciation	$ 2,800.00
Increase in inventory of livestock	$15,000.00
Feeder calf sales	$41,000.00
Total farm cash operating expenses	$57,000.00
Machinery depreciation	$12,000.00
Decrease in crop inventory	$ 4,000.00

PROFIT-LOSS STATEMENT

Cash Farm Income _____

Inventory Change _____

Total Revenue _____

Total Cash Expenses _____

Depreciation _____

Total Expenses _____

Net Farm Income _____

Conclusions

1. What is the total revenue?

2. What are the total expenses?

3. What is the net farm income?

4. What efficiency factors can be determined using the information from a profit-loss statement?